Reincarnation

ALSO BY THE AUTHOR

The Unity Principle: The Link between Science and Spirituality
The Nonlocal Universe: Why Science Validates the Spiritual Worldview

Praise for *Reincarnation*

"Steven Richheimer's book *Reincarnation: Science of the Afterlife* is a splendid introduction to this fascinating topic. He provides his readers a cross-cultural historical survey, a review of the best-designed scientific studies, and a description of how data on past-lives enriches the field of consciousness studies. Readers new to the topic will find no better introduction, and those well-versed on the subject will come across insights that are original as well as inspirational."

— Stanley Krippner, PhD, psychologist, parapsychologist, and Professor of Psychology at Saybrook University, Pasadena, CA. Co-editor, *Varieties of Anomalous Experience: Examining The Scientific Evidence*

"This is a fascinating and engaging book which is of great value for informing the general public about the important knowledge that the dominant materialist paradigm can, indeed, be meaningfully questioned. The spiritual worldview that he outlines, discernible in a variety of religious traditions and philosophies from around the globe and across history, is something which humanity greatly needs in an era of conflict amongst increasingly polarized ideologies. It is a work I hope to be citing a great deal in the coming months and years."

— Jeffery D. Long, PhD, professor of Religion and Asian Studies, Elizabethtown College, Elizabethtown, PA. *Author of A Vision for Hinduism, Jainism: An Introduction*, and the *Historical Dictionary of Hinduism*. Editor of Perspectives on *Reincarnation: Hindu, Christian, and Scientific*.

"In this book Dr. Richheimer presents the trailblazing scientific research on reincarnation conducted in modern times by psychiatrists Ian Stevenson and Jim Tucker—thousands of well-documented cases where children have told stories of what appears to be former lives, sometimes also presenting accompanying birth-marks, which gives an additional weight to their stories. He explains how ESP, mystical, out-of-body, and near-death experiences, as well as related phenomena may be explained in a meaningful way within the framework of a spiritual worldview—an explanation which, he argues, is supported by the intriguing ideas of quantum mechanics about nonlocality. He discusses some traditional practices—yoga, meditation etc.—which through the centuries have proven helpful for those wandering 'the spiritual path' with the ultimate goal of unification, through love, with the Divine. All in all, I think it would be difficult to find a more concentrated and intelligent introduction to these complex themes than Richheimer's book."

— Terje G. Simonsen, PhD, author of *A Short History of (Nearly) Everything Paranormal: Our Secret Powers - Telepathy, Clairvoyance and Precognition*

"First the author explores the history of reincarnation, which is a central tenet of many ancient cultures and religions. In the second part, he reviews the large body of scientific evidence that some children have accurate memories of a previous life, which is bolstered by substantial evidence that birthmarks, birth defects, and elements of personality are passed from one incarnation to the next. In part three, we see evidence that indicates that the mind is not physical and hence can survive death of the body. In addition, the evidence from quantum mechanics demonstrates that nonlocality applies not just to the mind but to physical reality as well. Finally, the author shows us that reincarnation is relevant if the true purpose of life is to achieve union with the Supreme Consciousness. This is achieved through moral behavior and some sort of intuitional practice such as meditation. I recommend the book highly to anyone who is interested in seeking knowledge of spirituality and the afterlife."

— Dada Maheshvarananda, author of *After Capitalism: Prout's Vision for a New World*

Reincarnation

Science of the Afterlife

Steven L. Richheimer, Ph.D.
InnerWorld Publications
San Germán, Puerto Rico
www.innerworldpublications.com

Copyright © 2019 by Steven L. Richheimer

All rights reserved under International and Pan-American Copyright Conventions

Published in the United States by InnerWorld Publications, PO Box 1613, San Germán, Puerto Rico, 00683

Library of Congress Control Number: 2019904942

ISBN: 9781881717737

Cover Design: Rodrigo Adolfo

All rights reserved. This book, or parts thereof, may not be reproduced in any form or by any means, electronic or mechanical, including photocopying, recording, or by any information storage or retrieval system, without permission of the publisher except for brief quotations.

For my beloved guru, Baba. He is my guiding light.

ACKNOWLEDGMENTS

I would like to thank my wife and spiritual companion, Jeanne P. Richheimer, for her help in doing the illustrations. I am also deeply indebted to Devashish Donald Acosta for his help in editing and laying out the final manuscript for this book.

Contents

Introduction	1
PART I: THE BACKGROUND	5
1. Reincarnation in the East	7
2. Reincarnation in the Ancient Egyptian and Greek Religions	13
3. Reincarnation in Judaism	17
4. Reincarnation in Christianity	21
5. The Law of Karma	36
PART II: THE EVIDENCE	47
6. The Studies of Ian Stevenson	52
7. The Work of Jim Tucker	58
8. They Remember their Previous Life	62
9. Prodigies and Geniuses	71
10. The Biology of Reincarnation	76
PART III: THE MIND	87
11. Unity of Self and Memories	91
12. Mind and Body	99
13. Mystical Experience	104
14. Out-of-Body and Near-Death Experiences	114

15. ESP	123
16. The Nonlocality of Quantum Mechanics	140
PART IV: THE PATH	151
17. The History of Unity	155
18. The Importance of Moral Behavior	160
19. The End of Rebirth	168
Notes	178
Index	188

Introduction

REINCARNATION IS DEFINED AS the transmigration of an individual's soul from one body to a new body following death. The word "soul" has many different meanings in different religions and cultures. Throughout this book, I will use it to mean our unit mind—the part of our being that survives death and in theory carries the memories, personality, and karma from previous lives to our present life. The self as opposed to the soul may be termed the "unit consciousness." It is the witness of the mind. Eastern religions refer to the individual self as the *atman* and the Supreme Self or witness of all creation as the *Paramatman*. This Sanskrit term is synonymous with Cosmic Consciousness or God.[1]

The concept of reincarnation is closely entwined with the idea of karma. The law of karma simply states, "we reap what we sow." Our karma is the sum of all our past actions, both good and bad, that we have retained in our unit mind but in a sense have not yet experienced or "burned." We are born into circumstances that allow us to experience life in such a manner that we can best burn through our remaining karma and progress toward the ultimate goal of union with God. In other words, our karma helps us learn important life lessons, including how some actions bring us happiness and others result in pain. Through this process, we grow in wisdom and move forward on the path to spiritual union.

In the first part of this book, we will explore the history of reincarnation and find that it is not only a central tenet of all the major Eastern religions but is also found in many other ancient and modern religions and cultures. For example, reincarnation was a tenet of the ancient Egyptians, Greeks, Celtics of Great Britain, Vikings, folk religions of Africa, Australia, East Asia, Siberia, shamanism of South America, and the indigenous Oceanic peoples. In addition, Native Americans have a long history of belief in reincarnation and today the Iñupiat people of northwest Alaska, the Aleuts of the Aleutian Islands, and the Tlingit Indians of Canada maintain this belief. Interestingly, the Tlingits believe that souls tend to return to their own families.

In addition, Judaism, Islam, and Christianity have deep ties to this doctrine; and to this day, the Kabbalist and Hasidic sects of Judaism and the Sufi, Ismaili, and Druze sects of Islam accept and teach the doctrine. During the first four to five hundred years of its existence, Christianity had a strong connection with reincarnation as evidenced by the beliefs of Gnostic Christians, the teachings of Origen, and other Neoplatonists. Some have argued that Jesus taught reincarnation but that the Church misinterpreted his teachings. As we will see, the belief in reincarnation has been around since ancient times, and although it is the accepted doctrine of most Eastern religions, it has survived as a heretical or out-of-the-mainstream belief within Judaism, Christianity, and Islam.

Today belief in reincarnation is common among the world's population; by some estimates the number of people who believe in reincarnation exceed the number who reject the idea or have no knowledge of it. Recent surveys showed that 20 percent of Americans believed in reincarnation and 26 percent of Canadians, while for Europeans the number was close to 30 percent.[2] The number of Westerners that believe in reincarnation might seem surprising considering the fact that it is not a matter of doctrine or faith. In fact, most Westerners consider reincarnation as esoteric, unrealistic, and inhuman—a doctrine originating in India that says a human can be reborn as an animal and has been used to justify casteism. Given these negatives concerning reincarnation we might conclude that for Westerners their belief in reincarnation springs from an intuitive feeling that their individual existence does not begin with their birth nor end with their death. Today most Westerners have a hard time accepting the orthodox Christian story of an afterlife which teaches that at the End Time the dead of those who are to be saved will rise from their graves and their physical bodies reconstituted in a heavenly realm where they will no longer age or be subjected to illness. For some people who believe in an afterlife it may seem more reasonable that we suffer the consequences of our actions and return repeatedly to earth until we attain salvation.

It should come as no surprise that throughout history many well-known and influential people stated they believed in reincarnation. A list of some of these people includes Pythagoras, Socrates, Plato, Dante, Voltaire, Napoleon, Kant, Goethe, Nietzsche, Thoreau, Emerson, Blake, Franklin, Jung, and Tagore. More contemporary personalities include Thomas Edison, Winston Churchill, Jack London, William Wordsworth, Albert Schweitzer, Walt Whitman, Henry Ford, Mark Twain, General George S. Patton, Eleanor Roosevelt, Princess Diana, John Lennon, Shirley MacLaine, Richard Gere, Tina Turner, John Travolta, Eckhart Tolle, Wayne Dyer, Willie Nelson, and Oprah Winfrey.

In Part II of this book, I will present some of the evidence that indicates that both children and adults have accurate memories of a previous life. Secondly, there is substantial evidence that birthmarks, birth defects, and elements of personality are passed from one incarnation to the next. Cases of genius, the selective but extraordinary abilities of savants, and the recall of previous lives by hypnotic regression are also indicative of reincarnation.

Reincarnation would not be possible without the survival of the soul following physical death. In Part III of this book we will explore some of the evidence that indicates that the mind is not physical—a requirement for life after death. As much as neuroscientists today want to equate mind and brain, there is extensive evidence that contradicts this hypothesis. If we look at the facts with an open mind, we are forced to conclude that mind cannot be equated with electrochemical brain activity.

Biological evolution traces our physical existence back to single-celled organisms. Hence our physical body can be traced back at least four billion years to the first emergence of life on this planet. However, if we possess a soul when and how did it first arise? By equating the soul with unit mind, we can conclude that our soul began with the first emergence of life on the planet. This is because even primitive life forms express simple behavior consistent with the idea that they possess a unit mind. We will also explore why we need more than the passage of genes from one generation to the next to explain the evolution of species, animal instincts, and behavior.

Since the soul can transmigrate from animal to human, there is an almost unlimited supply of souls to account for the ever-increasing source of humans populating the earth today. Unfortunately, the opposite migration would be possible—i.e. from human to animal. This could occur if the mentality of a human descends to a subhuman level. Fortunately, such a regression would be temporary since like the movement of a very slow clock, the souls of animals are inevitably moving toward incarnation in human form.

Reincarnation would be meaningless if the true purpose and meaning of life was anything other than to achieve union with God. This goal of ultimate union is found in the Eastern religions, and in the mystical traditions of Judaism, Christianity, and Islam. In Part IV of this book we will delve into this topic in some detail.

PART I

THE BACKGROUND

1
Reincarnation in the East

I appear before a person according to his or her desires. His or her whole being will be filled with My being. All the jivas (units) of this universe are rushing toward Me, knowingly or unknowingly. This is the final secret of the universe.

—Lord Krishna, Bhagavad Gita

REINCARNATION IS A CENTRAL tenet in just about all the Eastern religions. Those would include Hinduism, Jainism, Sikhism, Buddhism, and Taoism. The first known expressions of religion and the concept of reincarnation go back at least fifteen thousand years with the earliest religious text known—the Rig Veda. At this time, the human society had not yet invented script and was still undeveloped. People led a simple life. While the Rig Veda does not address reincarnation specifically, this earliest Veda does reference the idea of rebirth in several passages. For example, a funeral verse states:

May your spirit return again, to perform pure acts for exercising strength, and to live long to see the sun.

The Tantra of Shiva

Approximately seven thousand years ago a great sage by the name of Shiva was born in what is today India. It was a turbulent, conflict-ridden time in India, and the indigenous people clashed with the Aryans, who were outsiders that were beginning their migrations into the subcontinent.[1] Shiva apparently possessed occult powers from his childhood and did not have a spiritual teacher or the need to perform intense meditation to attain his divine status like the Buddha who came later. Shiva taught in the tantric tradition, which stressed spiritual growth through meditation as opposed to the Vedic tradition that at the time relied heavily on ritualistic practices. The Tantra of Shiva was a collection of his wife Parvati's questions (*nigama*) and Shiva's answers (*agama*). He taught that self-knowledge and eventual liberation were attained by spiritual practice or meditation (*sadhana*). He taught that this universe originated from a Cosmic Entity and is maintained in the vast body of that Cosmic Entity; and every living creature will eventually have to merge again in that Entity. Shiva taught that every individual has passed through numerous life forms on their path of evolution to human form and then may have to pass through many human bodies on their way to union with the Cosmic Entity—that is, be reincarnated.

Shiva is sometimes considered the father of spirituality because of his monotheistic concept of one God and his vision that all things spring from and are manifestations of the Cosmic Entity. Although Shiva's greatest contribution to humankind was in the arena of self-knowledge and defining a path for attaining salvation, he also made significant contributions in the areas of music, dance, hand and body gestures (*mudras*), system of marriage, science of breath control, ethics, and medicine (Ayurveda).

Following his death Shiva became accepted as a divine personality and today he is worshipped as one of the principle incarnations of God in the Hindu pantheon.

Vedas and Upanishads

The Sanskrit word *veda* means "knowledge or wisdom." The Vedas are a large body of religious texts originating in ancient India. They were composed in Sanskrit and passed down orally for many years before being put in writing. The Vedas are considered the basic scriptures of Hinduism, and Hindus consider the Vedas to be inspired by God. Although the Vedas are the foundation for Hinduism, the Vedic literature of the Upanishads was the principle inspiration for the school of Hindu philosophy known as Vedanta (literally: end of the Vedas). The central spiritual concept of the Upanishads is that atman (soul, self) is the "seat" of the mind and that in order to prevent reincarnation (*samsara*) one must attain moksha (salvation, liberation, union with God). The Upanishads also teach the concepts of karma and *dharma* (the human desire to attain limitlessness). According to the Upanishads, it is the individual's atman or self that reincarnates until the atman attains unity with the *Paramatman* (Cosmic Consciousness). These fundamental spiritual ideas were shared with other Eastern religions. The Upanishads stand among the most important literature in the history of Indian religion and culture.

The Upanishads can be considered as the distillation of the spiritual ideas of the Vedas and were the basis for the transformation of Hinduism from a ritualistic religion to one that was primarily spiritual. This transformation was aided by the sage Shankara in the eighth century CE. At that time, Buddhism had become the dominant religion in India, and Shankara traveled throughout the country reuniting the fragmented Hindu religion under a systematic monistic philosophy (Advaita Vedanta) that was based upon the Upanishads. Now recognized as one of India's greatest mystics and philosophers, Shankara taught that dualism was an illusion (maya) and that everything was the manifestation of the One (Brahma); to know one's essence or atman, the core of oneself, is to know Brahma. He taught that the human mind could move in two directions—toward the finite or toward the infinite; toward *preya* (illusory happiness) or toward *shreya* (permanent bliss). Although

Shankara was a great advocate for knowledge, he admitted that of all the ways to attain salvation devotion was the greatest.

Taoism

The yin-yang symbol

Lao Tzu (~500 BCE) is traditionally regarded as the founder of Taoism and is credited with writing the Tao Te Ching, the fundamental scripture of Taoism. Taoism and Confucianism are the two most important social-spiritual traditions that originated in China. Tao means "the path or way of life" but also denotes the principle that is the source, pattern, and substance of everything that exists. Taoism teaches that to know God, you must know yourself and that this may take many lifetimes. Hence, reincarnation is a basic tenet of Taoism. Taoism describes all change as the interplay between the two complementary forces of yin and yang. Although these forces are considered opposite, they are complementary parts of the One. The yin-yang symbol represents these two complementary opposites within the whole. Yin and yang describe the introversal and extroversal duality inherent in nature, but they are like two sides of a coin—the coin representing the One. Truth lies in the One, and knowledge of the One is the only true knowledge.

Buddhism

By the fourth century BCE, Hinduism had degenerated to a religion of lifeless, ritualistic practices. This set the stage for the emergence of a revolutionary new religion founded by Gautama Buddha. The Buddha (~400 BCE) was born as Prince Siddhartha Gautama in what is now modern-day Nepal. The story of Siddhartha's life begins with an astrologer telling his father, the king, that his son would become either a world ruler or a great spiritual master. This led his father to try to insulate his son from the knowledge of the existence of sorrow and death—lest he desire to take a spiritual path. Siddhartha grew up as a pleasure-loving youth who eagerly enjoyed all the sensory pleasures that life could offer.

As the story goes, one day Siddhartha ventured from his luxurious palace and happened to see a feeble old man, hobbling with a cane. Having never witnessed any evidence that people aged, became sick, and died he demanded an explanation from his charioteer who admitted that eventually everyone succumbs to old age and dies. Next he met a monk who told him that the way to attain that which does not change with time is to renounce one's worldly possessions and the short-lived world of sensory pleasures and devote one's entire being to becoming enlightened. Shortly thereafter Siddhartha renounced his kingdom, family, and worldly possessions, and began his long and difficult journey to become the Buddha, the Enlightened One.

After attaining enlightenment, the Buddha acquired a large following inspired by his message of morality, compassion, and the Eight-Fold Path to enlightenment. These are: (1) proper philosophy, (2) proper determination, (3) proper speech, (4) proper occupation, (5) proper exercise, (6) proper work, (7) proper meditation, and (8) proper attainment of samadhi or union. He opposed the ritualistic religions that were popular at the time and denounced asceticism as unnecessary and the wrong way to attain enlightenment.

Buddha taught four principles called the Four Noble Truths. First, there is suffering, second there is a cause of suffering, third there is cessation of suffering, and finally there is a way to end suffering permanently. By emphasizing suffering as central to the human condition, he was speaking to people of that time who were accustomed to hardships and suffering in their lives. He could just as easily have formulated these truths around happiness since both are conditions of the mind, and it is only when the mind is suspended that both conditions disappear and there is the experience of indescribable bliss (ananda).

Buddha did not explicitly mention or deny the existence of Cosmic Consciousness or God. It is believed that Buddha refrained from answering questions about the existence of God because he knew that God's existence is beyond the scope of the human mind to understand and should not be a matter of doctrine. Instead of preaching about God, Buddha taught *Shunyavada*, the Doctrine of the Void. According to him, everything of this universe emanates from *shunya*, everything is maintained in *shunya*, and finally everything will merge in *shunya*.

Buddha strongly supported the doctrine of rebirth. He taught that our journey on this earth is a continuous succession of birth, life, death, and rebirth until we ultimately attain liberation (*mahanirvana*). Buddha did not talk about the existence of a unit consciousness or atman in man. However, since he taught that man must be reborn in human form, it follows that man must possess a unit consciousness, for without it rebirth would be meaningless.

2

Reincarnation in the Ancient Egyptian and Greek Religions

The end of life is to be like God, and the soul following God will be like Him.

—Socrates

Ancient Egyptian ideas about the afterlife

THERE ARE A NUMBER of similarities between the Egyptian beliefs in the nature of the human soul and afterlife and those of Eastern thought. Hence, it is likely that some of the early Egyptian ideas of the afterlife were influenced by Eastern mysticism. One of the similarities was the Egyptian belief that humans possessed a *ka*, or life-force, which left the body at the time of death. Similar to the *prana* of yogic philosophy, the *ka* received its sustenance from food, air, and water and would survive death. Hence, the *ka* could be identified with the soul. In addition, Egyptians believed there was a divine witness of creation, the spiritual *Ka* that paralleled the *Paramatman* of Hinduism. Rebirth would continue until the individual's *ka* united with the spiritual *Ka*. The propensity of the ego to associate with the lower planes of existence would cause the soul to associate with a new body and be reincarnated.

The Egyptian ideas of the afterlife evolved into an elaborate set of beliefs that differed from Eastern mysticism in several respects—principally in their belief that individuals also had a *ba* that unlike the *ka* remained attached to the body after death. Thus if the corpse was mummified it would prevent decay of the body and prevent the soul from reincarnating in at new body. During the period between death and rebirth the soul had the freedom to adventure throughout the material and spiritual realms nurtured by the psycho-spiritual emanations of food and other material objects left in its tomb.

Orphism

Orphism is the name given to a set of religious beliefs and practices originating in the ancient Hellenistic world. Beginning about the sixth century BCE, Orphism was a set of religious beliefs and practices associated with literature ascribed to the mythical poet Orpheus. It was perhaps the earliest purely Western philosophy about reincarnation and the mystical union of an individual with God.

Orphism characterized human souls as divine and immortal but doomed to live (for a period) in a "grievous circle" of successive bodily lives through transmigration of souls. It prescribed an ascetic way of life, which, together with secret initiation rites, was supposed to guarantee eventual release from the cycle of rebirth and final communion with God. Orphism was supposedly founded upon sacred teachings about the origin of the gods and human beings.

Other Greek teachers of reincarnation

Several well-known Greek philosophers such as Pythagoras, Socrates, and Plato taught reincarnation. The Greek word for reincarnation was "metempsychosis," which derives from *meta* (change) and *empsykhoun* (to put a soul into).

Pythagoras (~570-495 BCE) was a student of Orphism. He is best known for the Pythagorean Theorem, but he is also known for his discovery that music had mathematical foundations and regular intervals. Pythagoras established a school in Croton, Italy around 530 BCE. He taught that the soul of man survives bodily death and migrates into a new body until it attains mystical union with God. He stated that he had experienced all forms of life, including "a boy, a girl, a bush, a bird, and a mute fish in the sea." In fact, the name Pythagoras meant "he who remembers his incarnations." Many of Pythagoras's ideas about the immortality of the soul, marriage as a spiritual tie, diet, and the importance of performing spiritual practices mirrored Hindu beliefs and support the theory that Pythagoras traveled to India during his lifetime.

For example, Pythagoras believed that a sentient diet that was free of dead and decaying animal flesh was good for the body and mind. In addition, he advocated fasting, which would aid in the purification of the body that was needed to ascend to the higher realms.

Pythagorean ideas exercised a marked influence on Plato and through him, on all of Western philosophy. Much of the surviving information about Pythagoras originated with Aristotle and the philosophers of the Peripatetic school.

Plato (~427-347 BCE) founded the Academy in Athens, the first institution of higher learning in the Western world, and unlike nearly all of his philosophical contemporaries, his entire work is believed to have survived intact for over 2,400 years. Plato was a student of Socrates and both men argued that the soul of man was superior to the body, as it was divine and eternal. Furthermore, the soul passes from one body to the next until it finally attains the realization that it is divine, and after it attains its divine home there is no compulsion to incarnate again. Hence, Plato strongly endorsed reincarnation, arguing that the soul needs more than one incarnation to acquire the experiences and lessons required before it can attain perfection. He believed that once a new life has been chosen by the soul and rebirth was imminent, the soul must drink from the river of forgetfulness, so that it is born without the knowledge of the eternal nature of the soul and of all previous lifetimes.

Interestingly, Plato provides one of the earliest accounts of a near-death experience. He tells the story of a soldier named Er who was fatally wounded in battle and being thought dead was placed on a funeral pyre, only to awaken and describe in detail his experience. Er described leaving his body and traveling to what he described as the afterlife. There he witnessed his past deeds and felt that the final destination of the soul was determined based on those deeds. He met with loved ones and friends long dead and experienced the great beauty and brilliance of a place "filled with a great radiant light." Then he described being drawn back to his body and awakening in pain but having the knowledge that death was not final but simply a transition. Plato's story of Er is primarily a lesson in reincarnation since it reinforced his argument that the soul survives death of the body, only to return in a new body in order to continue its journey to perfection.

Plato's allegory of the cave in his *Republic* similarly reflects his understanding that our everyday experience of reality is but a "shadow" of a "higher" reality. Chained inside a cave, looking at a wall dancing with shadowy figures from a candle, residents take their illusions to be reality. These prisoners deem reality to be nothing but the shadows of the actual objects. However, if a prisoner is freed and, climbing out of the cave with dazzled eyes, discovers the blazing sun and the true world that it floods with light, such a person would certainly describe all he saw before as a cheap illusion. However, he would have a difficult time explaining this experience to any of his co-inmates when he returned to the cave. They would surely laugh and mock him when he tried to convince them that their experience of reality was far short of what he had experienced.

Without doubt, Greek philosophy, particularly that of Plato (Platonism), had a profound influence on the early Christian Church, and furthermore it is likely that Jesus spoke Greek and was familiar with these ideas. Although Christianity held on to the concept of judgment and a heaven and hell, Plato's concept of reincarnation with forgetfulness of our past lives was eventually purged from Church theology.

3

Reincarnation in Judaism

Why are there evildoers who are well off and righteous who suffer evil? Because the righteous man was an evildoer in the past and is now being punished…not necessarily in his present lifetime.

—Sefer ha-Bahir (Kabbalistic text)

THE BELIEF IN REINCARNATION existed among Jewish mystics in the Ancient World. It stands as one of several different explanations given of the afterlife and belief in an immortal soul. In modern Judaism, reincarnation is an esoteric belief seen only within several mystical streams of Judaism. It is not an essential tenet of traditional Judaism and is not specifically mentioned in traditional classical sources such as the Hebrew Bible or the classical rabbinic works (Mishnah and Talmud). However, reincarnation may be the key to an understanding of many biblical passages.

There is evidence to suggest that Jews at the time of Jesus accepted the Greek doctrine of preexistence of the soul and reincarnation as a matter of course. The Jewish historian, Flavius Josephus (37–100 CE), wrote that there were three sects of Jews during that era, the Sadducees,

Essenes, and Pharisees. Josephus wrote that two of the sects, the Essenes and the Pharisees, believed in reincarnation. Josephus wrote:

> The Pharisees believe that souls have an immortal vigor in them and that the virtuous shall have power to revive and live again: on account of which doctrines they are able greatly to persuade the body of people.

Josephus records that the Essenes of the Dead Sea Scrolls lived "the same kind of life" as the followers of Pythagoras, who taught reincarnation. According to Josephus, the Essenes believed that the soul is both immortal and preexistent, necessary tenets for belief in reincarnation.

Josephus himself served as a soldier and once rallied his men to fight by citing the doctrine of reincarnation. Josephus said to his men:

> Do ye not remember that all pure spirits when they depart out of this life obtain a most holy place in heaven, from whence, in the revolutions of ages, they are again sent into pure bodies.

The Old Testament has several references to reincarnation. For example, in Jeremiah 1:5 God tells Jeremiah that before he was formed in the womb He knew him and appointed him a prophet. This implies that the soul of Jeremiah preexisted his birth in the sixth century BCE and had qualities appropriate for his becoming a prophet. In the Wisdom of Solomon, the author states the he was endowed with a good spirit and entered an undefiled body—implying that his soul preexisted his birth.[1] In the Book of Job it is said:

> Behold, all these things does God do—twice, even three times with a man—to bring his soul back from the pit that he may be enlightened with the light of the living.[2]

In other words, God will allow a person to come back to this world from "the pit" (purgatory) a multitude of times, until he becomes enlightened. This is clearly a reference to reincarnation as a process that may be required before the soul can attain perfection or union with God.

Finally, Malachi tells us that the spirit (soul) of Elijah will return to earth before the Messiah.[3]

The Jewish philosopher and mystic Philo was a contemporary of Jesus who lived in Alexandria, Egypt. His greatest contribution to Judaism at this time was the harmonization of Jewish scripture with Greek philosophy. He wrote in the allegorical tradition of the time and argued that many of the people and events in the Old Testament were symbolic. For example, he wrote that Adam and Eve were not actual people but represented mind (Adam) and body (Eve). Some of the early fathers of the Christian Church, namely Clement of Alexandria, Origen, and Ambrose studied his writings and incorporated his ideas into the doctrine of the growing Church in the second through fifth centuries. After his death, some Christians claimed that Philo had converted to Christianity and he was even referred to as "Bishop Philo" in Byzantine manuscripts.[4] Many of the concepts he introduced, such as the "Logos" or the "Word" as God's creative principle, may have influenced the early Christian authors of the New Testament.

Philo believed that the soul was preexistent and would need to reincarnate until the goal of communion (union) with God was achieved. His philosophy on life after death mirrored that of Plato, Pythagoras, and other Orphic philosophers. He is but one example of the Hellenistic influence on Judaism from the time of Alexander until the early part of the Christian era. There is little doubt that the Jewish leaders of Judea and even Jesus himself would have been familiar with the philosophy of Philo.

Rabbi Simeon bar Yochai is credited with composing the Zohar, a classic Kabbalistic text in the second century CE. The Kabbalistic movement focused on the hidden, mystical wisdom of the Jewish faith. The Zohar was later edited and first published by Rabbi Moses de Leon in 1280. Below are two representative passages from the Zohar, regarding reincarnation:

All souls are subject to the trials of transmigration (reincarnation); and men do not know the designs of the Most High with regard to them; they know not how they are being at all times judged, both before coming into this world and when they leave it. They do not know how many transmigrations and mysterious trials they must undergo.

> Souls must reenter the absolute substance whence they have emerged. But to accomplish this end they must develop all the perfections, the germ of which is planted in them; and if they have not fulfilled this condition during one life, they must commence another, a third, and so forth, until they have acquired the condition which fits them for reunion with God.

Reincarnation in the Kabbalah is known as *gilgul*. The term means "the judgment of the revolutions of the souls." It is also a belief in Hasidic Judaism, which regards the Zohar and Kabbalah as sacred and authoritative. In Kabbalah, any higher spiritual truth is seen to be reflected in lower forms in this physical world. This is because the divine life force for this realm first descends through the chain of higher realms.

4

Reincarnation in Christianity

Once we are clear that Jesus did teach reincarnation, most of Christian theology must be reexamined and rewritten in light of this new truth. We must reexamine the 'plan of salvation,' the relationship between karma and grace, and the far-reaching implications of the spiritual nature of humankind.

—Herbert Bruce Puryear[1]

THE TITLE OF THIS chapter would not make much sense to most Christians living in the Western world. They might rightly ask the question: hasn't the Church denied that the soul preexists birth and the doctrine of reincarnation since its inception? The answer would be that they are partially correct—the Church has denied the possibility of reincarnation for the last fifteen hundred years. Prior to this time, it was a topic that was not only hotly debated but also believed by many Christians. It can even be argued that Jesus grew up during a time when reincarnation was accepted by most Jews, and that he actually taught reincarnation.[2] However, subsequent public and clerical "authorities" chose to interpret and even distort his message in such a way that it denies reincarnation for what could be called "political" purposes.

Reincarnation during the time of Jesus

We have seen that the Jews at the time of Jesus accepted the Greek doctrine of preexistence of the soul and reincarnation as a matter of fact. Two of the three Jewish sects in the first century CE—the Essenes, and Pharisees believed in reincarnation. Hence it is no surprise that we hear in the Gospels of Mark, Matthew, and Luke that there was speculation among Jesus's followers that he was the reincarnation of Elijah or Jeremiah, or one of the prophets.[3]

If most Jews during the time of Jesus believed in the immortality of the soul and that more than one lifetime on earth might be needed to attain salvation, it is quite likely that Jesus believed in and actually taught reincarnation. There is considerable evidence for this conclusion based on the words attributed to him in the four accepted Gospels and in the "unofficial" Gnostic Gospels.

The return of Elijah

Possibly the strongest argument that Jesus taught reincarnation is his identification of John the Baptist as the incarnation of Elijah. In the Old Testament, Malachi proclaimed that before the Messiah the prophet Elijah would return to earth.[4]

According to Malachi, the reason for Elijah's return would be to "turn the hearts" of fathers and their children to each other. In other words, the goal would be reconciliation. In the New Testament, Jesus reveals that John the Baptist was the fulfillment of Malachi's prophecy:

> All the prophets and the law prophesied until John. And if you are willing to receive it, he is Elijah who is to come.[5]

This fulfillment is also mentioned by Mark and Luke.[6]
Specifically related to the prediction of Malachi, Jesus's disciples asked him why do the scribes say that Elijah must come first, and Jesus

answered them that indeed, Elijah is coming first and will restore all things, and that Elijah had already come but, tragically, he was not recognized and had been killed. His disciples then understood that he was referring to John the Baptist. Jesus then predicted he would likewise die at the hands of his enemies.[7]

A brief look at the ministry of John the Baptist reveals many similarities between his life and that of Elijah. First, God predicted John's work would be in the spirit and power of Elijah.[8] Second, he dressed like Elijah.[9] Third, like Elijah, John the Baptist preached in the wilderness.[10] Fourth, both men preached a message of repentance. Fifth, both men withstood kings and had high-profile enemies.[11] Sixth, just as Jesus proclaimed those who live by the sword die by the sword. Elijah ordered the execution of 450 priests of Baal by the sword and likely wielded the sword himself.[12] John suffered a similar fate when he was beheaded.

Some argue that John the Baptist was not Elijah reincarnated because he denied being Elijah and that in John it is said that he was only born in the "spirit" of the prophet. However, there are two reasons to doubt John's denial of being Elijah. First, it was written that Elijah never died.[13] When John denied being Elijah, he could have been countering the idea that he was the actual Elijah who had been taken to heaven. Secondly, John's words could indicate a difference between his view of himself and Jesus's view of him. John may not have seen himself as the fulfillment of Malachi's prophecy. He was a simple and humble man giving an honest opinion of himself. However, there is no doubt that Jesus credited John as the fulfillment of Malachi's prophecy regarding the return of Elijah.

Nor is there any doubt that Jesus clearly and unequivocally identified John the Baptist as the incarnation of Elijah in several passages of the New Testament. There is no ambiguity in Jesus's proclamation that John was the reincarnated soul of Elijah. Any modern interpretation suggesting that John was born only in the "spirit" or "essence" of Elijah is contrary to what Jesus said and is simply an attempt to deny the clear intent of the Bible in reaffirming the doctrine of reincarnation—a doctrine that was widely held and accepted for the first three to four-hundred years of Christian thought.

Other New Testament passages implying reincarnation

Jesus stated, "Is it not written in your law, I said ye are gods?"[14] He also said, "I am in my Father, and ye in me, and I in you."[15] Clearly, Jesus was indicating in these passages that the soul of man was part of God and as such was immortal and preexistent to birth.

In John 3:3 Jesus has a conversation with the Pharisee leader Nicodemus telling him that in order to enter the Kingdom of God one must be born again. Nicodemus either misunderstands this statement or thinks it refers to the prophesy concerning the return of Elijah, for he asks Jesus whether such a soul would have to suffer the indignity of being born of a mother or could he return as an adult. Jesus responds that one must be born of water (a physical body) and spirit before he can enter the Kingdom of God.[16] The context is that Jesus is talking about the process of resurrection—being born of the flesh and being born of the Spirit. These are two similar yet different processes. From these verses, the case can be made that Jesus taught the concept of resurrection as being physical rebirth (reincarnation) as well as spiritual rebirth (union with God). The orthodox view of these passages is that Jesus was only referring to spiritual resurrection, but from the perspective that Jesus believed in and taught reincarnation the context of the "born again" passages in John takes on a different meaning.

When Jesus walks by a man identified as blind from birth his disciples ask him whether this man sinned or his parents.[17] The latter explanation is consistent with the commonly accepted tenet of Judaism that people suffer for the sins of their parents. However, the first explanation would have to come from the general perspective that this man is suffering from sins committed in a previous incarnation. Clearly, Jesus's apostles considered reincarnation as a likely explanation for this man's blindness. Jesus answers that neither explanation is correct because a third possibility exists that the man took birth with this handicap so that he might experience and be an example of God's healing grace. Jesus's answer does not deny reincarnation—only that there is another possibility—he took birth for a special purpose. By answering this question in this broader tone, Jesus was reminding his disciples that one should not judge other people's suffering by concluding that they must have done something in the past to deserve their current misfortune. Belief in reincarnation should never be used as an excuse for withholding one's empathy and service to others less fortunate than oneself.

The law of karma implies reincarnation

Both the Old and New Testaments have numerous references to the idea that you reap what you sow—the law of karma. For example, in Genesis we learn that whomever sheds man's blood will have his blood shed,[18] and in Exodus it is written that if a man strikes another so that he dies, he will surely die, and that there will be a life for a life, an eye for an eye, tooth for a tooth, etc.[19]

In the New Testament, it is said "one reaps what one sows" and "those that live by the sword, die by the sword."[20] Even the Golden Rule implies that we should perform good deeds since the fruits of our actions will reflect back on us. Paul also describes the law of karma in his letter to Galatians.

> For every person will bear his own load...Do not be deceived: God cannot be mocked. A man reaps what he sows. For the one who sows to his own flesh will reap corruption from his flesh, but the one who sows to the Spirit will reap from the Spirit immortality.[21]

The biblical references to the idea that you reap what you sow, like begets like, or what goes around comes around, only makes sense in the context of reincarnation. Without the opportunity to experience the reactions to our actions in our present or future body there cannot be any fulfilment of our destiny to become fully divine and attain spiritual union with God and hence immortality. Certainly, a loving and merciful God would provide a means for all his children to attain salvation—even those born into a less advantageous condition and who might never have the opportunity to learn about Jesus Christ. If one is to suffer from one's misconceived actions at a future time it makes no sense that one would be judged worthy of salvation after just a single trip on earth.

Gnostic Christianity and reincarnation

The Nag Hammadi texts (also known as the Gnostic Gospels) are a collection of early Christian and Gnostic texts discovered near the northern Egyptian town of Nag Hammadi in 1945. Undisturbed since their concealment almost two thousand years ago, these manuscripts of Christian mysticism rank in importance with the Dead Sea Scrolls. These writings affirmed the existence of the doctrine of reincarnation being taught among the early Jews and Christians. The discovery in 1945 yielded writings that included some long lost gospels, some of which were written earlier than the known gospels of Matthew, Mark, Luke, and John. For example, in the Secret Book of John, written no later than 185 CE, reincarnation is directly discussed as the alternative to salvation.

> All people have drunk the water of forgetfulness and exist in a state of ignorance. Some are able to overcome ignorance through the Spirit of life that descends upon them. These souls will be saved and will become perfect, that is, escape the round of rebirth.

The Christian Gnostics emphasized spiritual knowledge rather than blind faith as the road to salvation. The Gnostics claimed to possess secret knowledge (i.e., "gnosis" in Greek) concerning the hidden meaning of the "resurrection." This was a part of the secret teachings of Jesus handed down to them by the apostles. This special knowledge was restricted to people who were qualified and initiated. In contrast, the very term "Catholic" means "universal," implying that anyone could become a member of the Church by adhering to the public teachings of faith and rituals. The Gnostic Christians were harsh critics of the orthodox Church. The Gnostics accused the Church of watering down the gospel in order to popularize it for the masses. They emphasized that people have three parts, the body, soul, and divine spark or spirit that connects them with God. The true goal of human existence was to nurture this divine part of their being and ultimately to become one with God. The Gnostics followed the cult of mysticism and held that knowledge of self was knowledge of God.

This secret gnosis emphasized spiritual resurrection (i.e. spiritual rebirth or union with God) and reincarnation as opposed to a resurrection defined as people sleeping in their graves until it was time for them to crawl out of their graves on the Last Day. Hence, Gnostic Christians held the view that if spiritual union with God was not attained in one lifetime, then the soul would be reincarnated until spiritual rebirth was attained.

Gnosticism reached its heyday in the second and third centuries and some estimates have put its number of adherents at that time as equal to orthodox Christians.[22] However, Gnosticism was not a unified movement, and a different person was selected to lead the group each time they met. There were also Jewish gnostics and Neoplatonist gnostics and other mystical groups that considered themselves gnostic, but Christian Gnosticism lacked a system of hierarchy or organization like in the orthodox Church. When the Emperor Constantine called the Nicaean Council, Gnostic Christians were not invited and this set the stage for the rejection of the Gnostic Gospels in favor of the canonical Gospels of Matthew, Mark, Luke, and John. Eventually the Gnostic Christians were labeled as heretics and mercilessly persecuted, executed, and wiped out.

Origen and the preexistence of the soul

Origen (185-254 CE) was born in Alexandria of Christian parents and was well educated in Greek and pagan literature, especially in Platonism—he even studied under the father of Neoplatonism, Ammonius Saccas. In his attempt to synthesize Platonism with Christian theology, he argued that the individual soul begins at the time of creation and therefore preexists birth, and that it continues to incarnate on earth until it obtains perfection.

He was a prolific writer and wrote more than two thousand treatises in multiple branches of theology. He is considered one of the most influential figures in early Christian theology and has been described as the greatest genius and theologian the early Church ever produced.

An example of Origen's argument for the preexistence of the soul and reincarnation was his commentary on the fate of the twins Jacob and Esau, who were grandsons of Abraham. In the Old Testament, it is written that God loved Jacob and hated Esau.[23] It is questionable whether God actually hated Esau, but Jacob was clearly favored and went on to father twelve sons that became the twelve tribes of Israel. Origen argued that God's favoritism could only be due to three possibilities. First, that Esau sinned while in the womb (a far-fetched proposition at best), secondly that God is unjust, or thirdly, that the boys earned their fates in previous lives. He argued that a loving and just God could only be describing the boys' fate while they were still in the womb due to their actions in previous lives.

Origen's arguments for the preexistence of the soul and reincarnation are found in many of his writings and reflected the prevailing belief among Gnostics, Platonists, Pythagoreans, and Neoplatonists at the time. God is inherently just and our past-life actions rather than the whims of God are responsible for our present condition in life. These ideas of Origen resulted in a controversy in the Church regarding preexistence of the soul and reincarnation that lasted no less than three centuries.

Origen's argument that each person is responsible for their fate in life went hand-in-hand with his idea that humans have free will. To him a just God would never create a world where the fate of some persons was sealed from the time of their birth—some to be saved, others to be condemned to damnation. According to Origen, each person has the opportunity to choose the path he or she takes in life—there is no such thing as a predetermined path to salvation. One has to earn this reward by actions in accord with Christ's message of love.

The concept of free will did not go well with the orthodox leaders of the Church, for it implied that salvation did not depend strictly on whether a person was baptized and regularly attended Church; instead it depended on their choices in this life and previous lives. Each soul has an opportunity to achieve salvation in this life, and if this failed, they might achieve it in a future incarnation. Hence the clergy were powerless to insure salvation—each person was free to make choices that would determine how they fared. As orthodoxy gained the upper hand in Church politics during the next two hundred years, the theology of Origen was increasingly criticized and his ideas about the soul, its origin, and its spiritual journey toward perfection were gradually looked upon as heretical. Origen's teachings were potentially quite damaging to the Church since the soul did not need the Church to achieve salvation.

The Nicaean Creed

The Roman emperor Constantine's conversion to Christianity in the fourth century set the stage for the official adoption of Christianity in the Roman Empire. However, at this time there was a significant "schism" in the Church regarding the divinity of Jesus. Many of the bishops at the time believed that Jesus was of the same divine nature as God, which implied that all humans were made of an "inferior substance" to God and the Son and could never attain the same status as Jesus or God. Another group of bishops led by Arius of Alexandria argued that Jesus was a begotten human who was subordinate to God and attained divinity upon his death. In other words, Arius and other bishops belonging to this school of thought rejected the contention that Jesus was of the same substance as God and taught that ordinary humans could also attain divine status like Jesus by following the path laid down by him.

This controversy was so divisive at the time that it threatened to undermine Christendom and harm what was an important unifying factor in the Roman Empire led by Emperor Constantine. As a result, Constantine decided to call a council in 325 in Nicaea to attempt to create a united and harmonious sentiment for this newly sanctioned religion of the Empire. The main contention to be decided was whether Jesus should be considered as part of God's creation and serve as a symbol for a human pathway to divinization, or whether Jesus was uncreated, an equal to God. In the latter case, man is created from a subordinate substance to Jesus (and God) and could only attain salvation through the Church and its law.

Constantine was both moderator and the ultimate authority in accepting what is known today as the Nicaean Creed. The Creed makes it clear that Jesus is not just the Son of God but also the one and only Son, constituted of the same substance as the Lord and different from that which constitutes humankind. Hence the Creed not only affirmed Jesus's divinity but also affirmed man's eternal separation from God. This ideological shift in Christian thinking continues until today and says that man's soul cannot be of the same "stuff" as God and it is therefore not of divine nature or eternal and can only come into existence along with the body.

Nicaea was the beginning of the end for the Christian concepts of preexistence of the soul, reincarnation, and salvation through union with God. Constantine's efforts to mold the Church into a state-sponsored religion compatible with his own image as an autocratic ruler was probably the first step in removing any semblance of mysticism from it. Gnostics were not invited to attend the Council and henceforth, with the blessing of the emperor, orthodoxy gained the upper hand. The doctrine that humans could not attain divine status like Jesus became increasingly established in the Church. This condemnation of the idea of the preexistence of the soul in Christian thought has continued until today. Some of the credit for this must also be given to a man who historians have recognized as one of the most influential figures in molding Western thought and Christianity—Augustine, bishop of Hippo.

Saint Augustine

Augustine (354-430) converted to Christianity at the age of thirty-three. Prior to his baptism, he practiced Manichaeism, which taught a dualistic cosmology where there is a struggle between the good, spiritual world of light, and the evil, material world of darkness. He possessed great intellect and formulated a Christian theology that was a synthesis of classical thought—essentially Platonism and Christian doctrine.

Augustine strongly rejected the reincarnation teachings of Plato and Origen. He argued that the soul is not preexistent and begins at the time of conception. Since some children are born into conditions of suffering, such as blindness or deformity, while others are born to ease due to no fault of their own, Augustine had to come up with an alternative explanation to that of the Origenists. His fallback argument was that such differences were due to the actions of one man—Adam—and the original sin that began with him. In other words, the cause of all human misery can be traced back to the mistake (fall) of one man. He even went so far as to argue that if a baby died before it was baptized it would suffer eternal damnation. By nature, man was wicked and the only way to overcome this was to access God's grace through the Church.

Augustine believed that Adam and Eve would have been immortal if they had not fallen from God's grace after eating the forbidden fruit of knowledge. After their fall, their offspring were condemned to experience suffering, old age, and death. According to Augustine, the role of Jesus was to restore some of God's grace for humankind. People would still have to undergo suffering and death while on earth, but by accepting Jesus into their life they could experience immortality and the absence of suffering by undergoing bodily resurrection. His belief system required that some souls were destined to come to Jesus and attain salvation while others were not so lucky and were destined to suffer eternal pain and punishment.

Augustine argued that creation was *ex nihilo*, or out of nothing. Creation *ex nihilo* was needed if God was to remain immaterial and perfect—pure spirit and not of this world, which is temporal and imperfect. Hence, God may be present and active in governing and organizing the universe but not "in" his creation except as Spirit. According to Augustine, the only presence of God in the material world is through the life of Jesus, or when he intervenes occasionally in the natural order by way of a miracle. To this day, this classical (orthodox) view of creation is accepted by the Church and assumes that God could create something separate from himself out of nothing.

The idea that there exists two fundamental "substances" in the universe—"God Stuff" and "out-of-nothing stuff"—is a decidedly dualistic philosophy. By demoting human beings to a lesser status than God,

Augustine's classical theism defined the orthodox position that human beings are both inferior and separate from God. The modern Church has had difficulty defending many of Augustine's ideas, but they have formed the basis for much Christian orthodoxy for the last 1500 years. Probably the most toxic elements of Augustinian theology is its implication that God's love for his created beings is not equally applied, and man is inherently inferior and separate from his Creator.

Elements of Augustine's dualistic Manichaeistic beliefs seem to have been carried over into his theology. For example, his zeal for heresy hunting took center stage in his mind and heretics were for the first time condemned to death in order that good could triumph over evil.

It is difficult to find any biblical authority for the Augustinian worldview. The Bible says "the Word was God" and everything that was made was made by the Word.[24] However, nowhere do the scriptures imply that the Word is nothing. On the contrary, the Bible teaches that the "Word became flesh" and "I am the vine; you are the branches."[25] Such scriptural passages contradict creation *ex nihilo*, and none can be found that support this concept, yet the dualistic theism of Augustine became the dominant theological position of Christianity.

The irrationality of the idea that the creation is not within God (i.e. God is not omnipresent) has led many Western thinkers to reject classical theism in favor of panentheism. Panentheism describes God as being both in the physical universe, i.e. "in" his creation, and transcendent to it. In this way, panentheism differs from pantheism, which is the belief that the creation is identical to God.

Eastern philosophers and prophets have always assumed that matter like everything else must be contained within the Divine Being. Hence, the Eastern religions are panentheistic: God, the Transcendental Entity, is one with his creation and continuously manifests as the creation.

The important Christian philosopher and theologian St. Thomas Aquinas saw creation as an ongoing process. Therefore, in some sense he proposed that God is continuously "in" his creation rather than an interested bystander. In other words, God is both transcendent and immanent. However, Aquinas believed God was perfect; his relationship with his created beings allowed for them to be autonomous and separate from him. Hence, Aquinas deviated in some respects from classical theism but did not fully embrace the concept of panentheism.

The end of reincarnation in Christianity

By the time of the rein of Emperor Justinian in the mid-sixth century, the Platonically inspired writings of Origen and other reincarnationists were officially rejected by the Church. Justinian summoned the Second Council of Constantinople in 553 to condemn Origen's teachings and the concept of reincarnation. He ordered that all his writings be burned. This action by Justinian set the stage for the complete acceptance of the orthodox versions of the Old and New Testaments. Interestingly history indicates that this Council was not attended by the Pope (Vigilius) and just a handful of Western bishops were present out of 152 total, and none from Italy. According to the historian Procopius, it was Justinian's wife Theodora whose ambition and hatred of the Origenist interpretation of Christianity was the real force behind the decisions of the Council. In any case, the Council officially decreed that Origen and his teachings were anathema—primarily because he taught preexistence of the soul and reincarnation.

Did Jesus study in the East?

The idea that Jesus traveled to India and Nepal during the missing years when the Bible tells us nothing about his life (ages twelve to thirty) is more than just speculation since considerable evidence exists to establish this as quite plausible. Elizabeth Clare Prophet details the evidence in her book *The Lost Years of Jesus*.[26] The principal evidence was in what has been called the Himis manuscript that a Russian writer by the name of Nicolas Notovich reportedly discovered in a monastery in Himis, Tibet.

This ancient text and other evidence garnered by Prophet reveal that Jesus spent seventeen years in the Orient from the age of thirteen to twenty-nine. In his travels to India, Nepal, and Tibet, he sought instruction from some of the great Hindu and Buddhist sages of the time with the objective of perfecting himself. He apparently made such a strong impression on religious leaders of the time that several ancient historians independently wrote stories about Jesus's travels in the East.

During such travels and studies in the East, Jesus would have become familiar with the concept of reincarnation, which is at the core of both religions. Prophet points out that there is evidence that the Vatican possesses some of the manuscripts and other evidence about Jesus's travels to the East. However, she suggests they are reluctant to release the evidence to the public since it would suggest that Jesus was not born a fully divine personality but went through a learning process to achieve his spiritual greatness, just like an ordinary human being.

Why did the Church ultimately reject reincarnation?

As mentioned earlier orthodox Christianity rejected reincarnation by the sixth century during the reign of Emperor Justinian. The doctrine of reincarnation was labeled anathema; any teachings of this doctrine were thereafter brutally suppressed.

Acceptance of the traditional theism of Augustine was one factor in this rejection. According to Augustine, human beings have a status separate and inferior to that of God. We are finite creatures tainted by sin instead of being children of the Infinite. Being separate or outside of God we can never attain divine perfection or union with him; therefore it is futile to return to earth to progress toward such perfection.

In addition, theologians may have felt that the doctrine of reincarnation would detract from the Church's message—salvation can be attained only through the grace of Jesus Christ. If humans were to return repeatedly to earth in order to attain salvation, then it would not be imperative to seek his divine grace in this lifetime.

Perhaps the main reason why reincarnation was rejected by orthodox Christianity is that it would inevitably lessen the control of Church leaders over their flock. The current system ensures that the clergy has the power to decide the conditions under which a soul will attain salvation and hence immortality in heaven—as opposed to condemnation to an eternity of suffering. There has been over a thousand years of history in which Church power has been used to the detriment of humankind. The doctrine of reincarnation would imply that one's salvation was ultimately in one's own hands and that God would be indwelling and omnipresent—self-realization could be attained without the help of a "middle man." The acceptance of such an ideology would greatly diminish the power of the Church.

5

The Law of Karma

Karma brings us ever back to rebirth, binds us to the wheel of births and deaths. Good Karma drags us back as relentlessly as bad, and the chain which is wrought out of our virtues holds as firmly and as closely as that forged from our vices.

—Annie Besant

SIMPLY STATED, THE LAW of karma is "you reap what you sow." Karma goes hand and hand with reincarnation—one does not make sense without the other. All the world's religions subscribe to the rule of karma even if they do not specifically endorse the doctrine of reincarnation. The idea that "what goes around comes around" is the reason one should follow the Golden Rule, since good actions beget good reactions and bad actions beget bad reactions. People who understand how actions, either good or bad, inevitably return to affect them in this life or in a future life have a reason to act morally. That is, they are aware that there is no escaping the consequences of their actions. Moral behavior is in one's own self-interest!

While Newton's third law describes the law of action and reaction for physical bodies—for every action, there is always an equal and opposite reaction—the law of karma expresses a similar idea in the psychic realm with the caveat that reactions to actions performed by a human being may be stored in the mind and experienced later. However, to understand the mechanism by which karma operates and why the law of karma implies reincarnation, we should first review some of the basic ideas that form the core of spiritual philosophy.

Spiritual philosophy—an overview

To someone who has never studied spiritual philosophy the concept of oneness or wholeness may be difficult to comprehend. After all, our experience and common sense tells us that every person and every object in this universe is different and separate. We do not feel that we have commonality with other people, to say nothing of a cow or a table. Yet the spiritual worldview proclaims that everything is a manifestation of the One, separateness and individuality are illusory, and the material universe we observe with our sense organs and manipulate with our motor organs is formed from Cosmic Consciousness. In other words, consciousness, not matter, is the fundamental "substance" of creation.

On face value, this proposition is no less plausible than the alternative, which we can call materialist ideology. Here subatomic particles make up atoms that come together to form molecules, and when sufficiently complex molecules form and assemble themselves in a particular way, simple living organisms arise that eventually develop mind and consciousness through an evolutionary process involving natural selection. This ontology fits right in with modern science and explains why everything is different. It is simple common sense that since the parts that make us up are separate and distinct, then we are separate and distinct from every other object in the universe. Any commonality depends on similarities in our chemical makeup—that we are made of the same atoms of carbon, nitrogen, oxygen, etc.—nothing more fundamental than this.

The big problem for spiritual ideology is to explain how something as nebulous as consciousness can be transformed into the material world with all of its distinct parts. To understand how this is possible we begin with the idea that creation is an ongoing process and is the internal psychic concoction of God, which for the purpose of this argument we will call Cosmic Consciousness (CC). This term is used because it is a better descriptor for the underlying characteristic of the Godhead. Consciousness implies awareness, witness, or observer. That quality that observes our mind is our unit consciousness (self or atman). However, the observer in us, our "I," might be compared to a ray of sunlight. It is of lesser quantity but no different in quality from the source—the sun. In other words, the individual self or atman is a part of CC (*Paramatman*), but it is directed outward observing our mind and body. If its direction could be reversed toward the source of our awareness then the illusion of a separate self would be removed and the true nature of our being revealed. A term used to describe this process is "self-realization."

When we conceive of a scene in our mind, such as a person riding a horse, no one else can witness it. However, for God his internal psychic creation can take on a physical form that appears as physical reality. To us the horse and rider in God's mind are real, but from the standpoint of the actual imaginer of the scene, CC, it would be no more real than the picture we created in our mind.

But the question remains how can CC, which is subtle and unqualified by nature, not even having the idea of its own existence, be transformed into the material world? The answer is that according to spiritual ideology, CC has within it a Cosmic Qualifying Principle (CQP). When this CQP first acts upon the unqualified CC, it produces the idea of "I exist" or "I am." Further qualification by the CQP on this "I exist" produces the "I do," and the "objective I" (cosmic mind-stuff), and these three qualities form Cosmic Mind. If we use the analogy of waves to picture this transformation of CC into Cosmic Mind then the CC would be a wave of infinite wavelength—a straight line. The "I exist" feeling would have a very slight curvature and thus a finite wavelength, and similarly the wavelength would be gradually shorter for the "I do" and "objective mind-stuff."

However, at this point the three components of Cosmic Mind remain purely subjective. It is as though the horse and rider have been created mentally in the Cosmic Mind but they have no physical reality because matter and energy have not yet been created. These are created in the next phase where the CQP gradually transforms the cosmic mind-stuff into space-time, gas, electromagnetic energy, liquid, and finally solid. These five fundamental factors are formed in a manner analogous to how cosmologists describe the formation of the universe following the Big Bang. The only difference is that spiritual ideology attempts to explain what precedes the Big Bang—the formation of Cosmic Mind from CC.

In the second or return phase of this cosmic cycle of creation, suitable conditions may exist on a planet or moon that allows consciousness to express itself within individual physical structures beginning with single-celled organisms. Similar to the Darwinist model of evolution, but under the influence and guidance of Cosmic Mind, the living organisms evolve leading to creatures with more and more complex mental and physical structures and ultimately to sentient or self-aware beings—what we call humans. Because the most subtle aspect of Cosmic Mind, the "I am" feeling, is fully reflected in the human mind, human beings have free will and are inexorably drawn toward the source of creation—CC. After many incarnations, each person will eventually merge or unite their individual mind (soul) with Cosmic Consciousness.

This is an abbreviated description of how spiritual ideology explains the creation of our universe and our place in it. The concept is central to most of the religious traditions of the East, including Hinduism, Vedanta, Buddhism, Tantra, yoga, Taoism, and Sufism. A common theme of these philosophies is that creation begins with CC and one's individual existence continues until it is merged or lost in the unqualified sea of CC. In other words, the creation is cyclical. It begins and ends in unqualified Cosmic Consciousness. Thus, spiritual ideology is described as "top-down" ontology as opposed to materialist ideology, which is "bottom-up." According to spiritual ideology, mind and matter are epiphenomena of consciousness.

The spirituality taught by the Christian Church is a mix of spiritual and materialist ideologies. God created the cosmos; the physical world is real—not illusory; humankind has a special status in the creation but we are radically different from God; we can attain salvation through the Church and God's grace but our salvation does not entail becoming one with God.

The mechanism for karma

Organisms evolve in the return phase of creation through the constant struggle for survival and the constant pressure of the CQP. Struggle causes the unit mind-stuff to become more subtle. This leads to greater and greater expression of subtler "I do," and "I exist" qualities of mind. The increased mental expansion is accompanied by greater physical complexity. In humans, the "I exist" portion of mind is fully expressed; ego and intellect have developed alongside self-awareness. Unlike less evolved animals, who are guided principally by instinct, humans are self-directed and thus experience the fruit of their actions—both good and bad.

According to this philosophy, every physical or mental action involves the unit's mind, but here mind is not to be equated with brain. The human mind reflects Cosmic Mind. Like Cosmic Mind it has an objective component called "unit mind-stuff." The horse and rider we create in our mind consists of this. The brain is not required to create this picture in our mind—only intention or "I do." Although mind requires brain to function through the sense and motor organs, the unit's mind-stuff can function independent of brain.

When the mind-stuff is vibrated by a mental or physical action, it retains an impression or reactive momentum (Sanskrit: *samskara*). Such a stored reactive momentum holds the potential for future action. When a stored reactive momentum is expressed, it is said to be "exhausted" or "burned." In other words the fruit of our action has ripened. One's karma is nothing but the totality of all these reactive momenta. The burning of a reactive momentum returns the mind to a "cleaner" state. An analogy might be to make a dent in a hollow rubber ball with your finger. The depression may last for a while, but if the ball is warmed with the hand, the dent can pop out, creating a symmetrical ball once again. In this example, the dent represents the reactive momentum. When the dent pops out, it is burned.

Reactive momenta result from mental and physical actions performed by the individual. They are stored as potential mental energy in the individual's unconscious mind, and their expression or burning is accompanied by the release of kinetic mental or physical energy equivalent to the mental energy or impression that created it. Although the amount of mental energy may be the same, the type and quality expressed will normally be different. For example, you do a kind deed for a stranger on a road by helping them change a flat tire. The reaction to such kindness will be stored in your mind and perhaps come back to you later in the form of help someone provides to you. However, it will probably not be help changing a flat tire. In a subtle, often unconscious way, the reaction teaches us that it is good to do good deeds and be nice to other people.

Similarly, a bad deed will ultimately bounce back, creating hardship, pain, unhappiness, or suffering. That mental disturbance reminds us either consciously or unconsciously that the action that created the reactive momentum was bad, and we will be less apt to repeat such action. This is the basic mechanism behind the law of karma. In other words, this law of action and reaction is a reward/punishment system by which we learn to be better, wiser, more self-realized, and move forward on the path to a perfect realization of self—ultimate unity with God.

Of course, we are free to direct our mind according to our whims and may choose to create both good and bad reactive momenta. However, we are not free as to when and how they are burned. The greater the mental vibration or intensity of an action, the more powerful will be the associated reaction and the stronger will be its effect when the reactive momentum is experienced. Strong desires create a strong potential for action. For example, a boy witnesses firefighters extricating a person from a badly damaged car and develops a strong desire to become a firefighter himself. After many years and extensive training, he may satisfy his urge to help people.

Actions performed without the "I do" of ego do not create reactive momenta. Therefore, the actions of others, actions performed unconsciously, and acts of God, such as floods or windstorms, do not create reactions in our mind directly. However, such experiences may create other reactions.

According to spiritual ideology, reincarnation is necessary in order for an individual to express unburned reactive momenta following their death. Cosmic Mind will find a new and suitable body for the bodiless mind in order that it can continue on the path of unification with CC. This could take place almost immediately or after many years. Once a suitable home for the bodiless mind is found, it will become associated with that fertilized ovum or zygote and eventually mature into a baby and be reborn. Thus, the minimum time between death and rebirth for humans is approximately nine months. The same process can be applied to animals, except that they are guided by instinct and not willful intent. Hence animals are incapable of performing actions that could be judged either good or bad. They do not suffer from the consequences of their actions and upon death simply take on a slightly more advanced physical and mental structure in accord with the inexorable movement of the creation cycle (evolution).

Since noncerebral memories retained by the bodiless mind can include memories of past events that took place in previous bodies, it is sometimes possible for a person to recall experiences from a previous life. Apparently, children are more likely than adults to recall memories of past lives—especially if they suffered a traumatic death. For children this can be a cause of anxiety, and in the West their parents will usually try to assure the child that these memories have no basis in reality. It is fortuitous that these memories of living in another body fade by the time they reach five to ten years of age.

For adults, remembrances of past lives may come out during dreams, deep meditation, or during hypnosis. A common clinical treatment for persons suffering from phobias or neurotic fears is hypnotic regression to a time when they first experienced an incident associated with the intense fear. Often, the traumatic experience they describe under hypnosis occurred in another body. Reliving this experience under hypnosis can cause a catharsis and help eliminate the phobia.

Contrary to popular belief about reincarnation, the vast majority of cases involve people recalling normal, unexciting lives. Except for medical treatment, the remembering of past lives is not recommended since it can create anxiety and uncertainty and can divert one's attention from the job at hand, which is to know one's self in the here and now.

The power to heal the sick may seem beneficial; however, it can involve taking on another person's karma, which can potentially rob the sick person of needed life lessons as well as adding that person's bad karma to the healer.

In the past, people were more inclined to avoid immoral behavior because of the fear of eternal damnation. This Church doctrine certainly helped keep people "in line" and discouraged them from actions that violated the Ten Commandments. However, today few people believe that they will be sentenced to eternal damnation in hell if they sin or fail to seek salvation through the Church. Belief in hell is at an all-time low. Fear as a basis for acting morally applies mainly to a fear of going to prison. The belief in the law of karma and reincarnation serves as a more logical rationale for acting morally, but unfortunately, it is not taught to most Westerners.

According to the spiritual worldview, we are God but do not realize it. Our life's purpose is to realize who we actually are. Therefore, actions that reinforce our connection with God reward us with happiness while actions that distance us are punished. Actions that draw us closer to self-realization (e.g. selfless service, spiritual practices, and meditation) produce the greatest happiness while actions that take us away from God produce reactive momenta that cause suffering. God has unconditional love for all his created beings, but the only way we can learn right action from wrong action and move forward on the path to union is if we are punished when we deny our divinity and rewarded when we affirm it. In other words, individuals who work to discover and develop their innate spiritual nature experience happiness and ultimately bliss. Those who take the path toward darkness and work against their true nature do not serve the purpose for which they were created and bring unhappiness and suffering upon themselves.

Some people believe that the karma gathered due to evil deeds can be compensated or neutralized by good actions. However, spiritual ideology suggests this cannot happen. All actions, whether good or evil, cause a deformity in the unit mind. In the process of the mind regaining a flawless form, the deformity is removed by an equal and opposite mental reaction. Hence the reactive momentum caused by an evil action cannot be removed by a good action. Every reactive momentum is independent of all others and one has to experience the consequences of good and bad actions separately. It is as though we made deposits into an account, which we must eventually withdraw.

Thus the results of good actions cannot help one evade the suffering caused by bad actions. At best, the mode of experiencing the reaction can be changed and the intensity of suffering can be reduced by slowing the speed at which reactions are experienced

It is obvious that all individuals carry a karmic burden. It could be called their "karmic bag." We take this bag from one lifetime to the next. The weight of this bag depends on how we acted in the past. The fastest way to empty our bag would be to go ahead and experience the consequences of our actions and not put any more reactions into the bag. Unfortunately, it is not enough just to perform good actions. This is why Buddha called good reactions "chains of gold." Whether the chains are gold or iron (bad reactions) they bind us to this world.

Why the law of karma makes sense

Church doctrine maintains that the soul comes into existence at the time of conception, and it is an accident of nature if an innocent child is born blind or without arms while another is born a prince. Because man is tainted by original sin, the fate of a child is left to chance because its ancestors committed a crime against God. According to the Old Testament, this crime was committed by Adam and Eve when they were tempted by Satan, a fallen angel, who spoke through a serpent and seduced them to disobey God's command not to eat fruit from the "Tree of Knowledge." The goal of the devil was to lead them away from the love of God into ignorance and evil. The suffering of innocents exists because our ancestors fell from the grace of God, committing what is termed "original sin."

The punishment for disobeying God's will and following a path of evil is eternal damnation in hell. Today few people believe this myth and it is undoubtedly one reason that Church attendance has declined worldwide. This doctrine is illogical if one believes in a God who loves all his created beings. It is not possible to believe that such a God would shower some with his love and grace and withhold it from others while sentencing some to a lifetime or even an eternity of suffering.

The question that the Church never really answers is why an all-loving, omniscient, omnipotent God would allow suffering and evil to exist in the first place.

The law of karma offers a simple and logical explanation for why some people seem to be born into a life of happiness and ease while others seem destined to suffering. Since reactive momenta remain in the unit mind after death, the bodiless mind, in order to find expression for its remaining karmic burden, takes on a new physical body. Hence we sometimes witness the phenomenon in which a child is born into a life of hardship and suffering due to no obvious fault of its own, while another child is born with the proverbial "silver spoon" in its mouth. Since we have no knowledge of a child's past lives, we remain mystified by such so-called accidents of birth. However, the spiritual worldview proclaims that everything in the universe is connected, and in reality, there are no accidents of birth. The reason some individuals seem to have a mountain of problems to overcome while others seem destined for happiness and success is that they brought it upon themselves by behavior in this life or in past lives. The fact that we do not remember our negative deeds from the past does not free us from having to reap their reactions. Humans are blessed that once the negative reactions are burned they are gone forever, and at a subtle mental level, they are reminded not to repeat the action that brought on the pain. Hence the burning of old reactive momenta and the production of new ones may go on for many lifetimes before the individual attains unity with God.

What about good and evil? Spiritual ideology claims that an action performed with good intention and knowledge brings one closer to the goal of becoming one with God and can be labeled good. Evil is just the opposite and is performed out of ignorance. In the simplest sense, good describes the movement from crude to subtle, or the process of identifying with the higher Self as opposed to the ego. The Sanskrit term *vidya* is used to describe this movement. Evil action may be termed *avidya*, or the movement toward ignorance or crudity. Such actions create negative reactive momenta, which lead inexorably to pain and suffering.

The evolution of unit mind occurs when the reflection of Cosmic Consciousness becomes clearer and greater in intensity. In this process, the mind becomes subtler and more expanded. This movement toward subtlety is accelerated by good actions such as selfless service and by focusing the mind on God as opposed to the crude material world. Because of the deep connection of the human mind with the limitless "I am" of Cosmic Mind, there is an inexorable tendency for the human psyche to be drawn toward the limitless CC, and every individual will eventually follow the path of knowledge and attain unification with God.

For the vast majority of people, progress in the social sphere has produced increasing awareness of the subtle connection they have with all of humanity. Artificial barriers such as nationality, race, religion, socioeconomic status, gender, etc. have slowly begun to dissolve as people have adopted a more universal and humanistic model of humankind. At the same time, moral attitudes have changed and grown. People have begun to develop a more family-like sentiment toward other members of society and increasingly have begun to treat others with the same love and respect that they might treat a brother or sister. Finally, when people become aware that the actions they perform will inevitably bounce back to affect them they experience moral growth and understanding, which must be encouraged if humankind is not to sink backward into the darkness of the past. There is no mistaking the fact that there are forces and dangerous ideologies that are both divisive and destructive to human values at work in the world today.

In the end, the greatest evil is the denial of our own humanity and to deny our connection to God. Such a mindset is no different from that of an animal, which is incapable of contemplating higher sentiments.

PART II

THE EVIDENCE

It may be of interest to the reader that some of the first "evidence" for reincarnation in the Western world came from a man whom few would dispute was the greatest American psychic—Edgar Cayce. Cayce was born in Kentucky in 1877 and received only an eighth-grade education before he was called to work on the family farm. He came from a devoted Christian family and was a devoted Christian himself. He would read the Bible cover to cover every year and regularly taught Sunday Bible School. It is therefore ironic that the Christian Church eventually labelled him as a false prophet and his work an inspiration of the devil.

At the age of twenty-three, after suffering from chronic laryngitis, Cayce sought help for his condition from a hypnotist. He was told to perform self-hypnosis in order to obtain a permanent cure. After entering a self-induced trance, he described in detail the ailment and its cure. This was his first psychic reading and over fourteen thousand were to follow before his death in 1945. He performed his readings while in a trance and after awakening had no recollection of what he said. Most of his readings were for people seeking medical diagnoses and treatments. All he required was the individual's name and location. He would give an accurate description of the illness or physical problem as though he had X-ray vision and then offer a treatment for the condition. The accuracy of his readings was astounding and the terminology he used was not that of an unschooled individual but of someone highly trained in medicine.

Edgar Cayce's first reading about the veracity of reincarnation came as a complete shock to him. When the text of answers to questions posed by Arthur Lammer in August of 1923 was read back to him he was very perplexed, since in the reading he had stated that, far from being a myth, the doctrine of reincarnation was a hard and fast truth. Cayce's dismay was based upon the fact that he was a devoted and orthodox Protestant and could not entertain the thought that any man was destined to return to earth in a new body to continue his path toward salvation. His faith flatly denied this possibility, and at first he wondered whether his psychic gift was being influenced by the devil. However, over time he began to understand that reincarnation and the law of karma were actually consistent with the teachings of Jesus and was the foundation for a spiritual philosophy that was both rational and consistent with the ideas of many mystical traditions.

By the time of his death, he had performed past-life readings for over twenty-five hundred people. He would often trace an individual's existence back thousands of years through numerous incarnations, pointing out how the person's current condition was influenced by past lives. Thus, Cayce strongly confirmed the law of karma and the reality of reincarnation.[1]

According to Cayce, talents and skills are never completely lost when a person completes an incarnation. Someone who has developed an ability in one life will be able to draw upon it in another. One may be born, for example, as a genius or prodigy in math because she developed this skill to a prodigious degree in a past incarnation.

Most of the evidence confirming reincarnation is in the personal accounts of individuals who have provided detailed and factually accurate descriptions of their previous life. Probably the majority of such accounts are of children who spontaneously describe their previous life to their parents in such detail that the person described could be identified and their life story checked against the facts provided by the child.

There are also numerous accounts of adults remembering previous lives—most often, under what is termed "regression hypnosis." Finally, we have cases suggestive of reincarnation, like those investigated by Ian Stevenson, where birthmarks and birth defects can be traced to circumstances of the death or to the personal history of the person identified as the predecessor of the reincarnated person.

The problem with much of this evidence is that it is anecdotal and depends on memories that cannot be expected to originate in the brain. Like all memories, we should expect there to be inaccuracies and gaps that may be filled in by everyday experience and knowledge—especially in the case of hypnosis because there is often a strong desire by the subject to fulfil the expectations and suggestions of the therapist.

An example of the problem is illustrated by arguably the most famous case of a past-life recall, that of Bridey Murphy. It is the story of an American named Virginia Tighe who under hypnosis talked in an Irish brogue and recounted memories of being a nineteenth-century Irish woman named Bridey Murphy. The result was a bestselling book, *The Search for Bridey Murphy*, written by her hypnotist, Morey Bernstein, in 1956. The book caused somewhat of a sensation at the time. However, Bernstein had not checked the facts related to him about the life of Murphy, and once the book became a bestseller, almost every detail was thoroughly checked by reporters, some of whom were sent to Ireland to track down the background of the elusive woman. It was then that some of the facts supplied by Tighe about her former life did not check out with historical records while other details did. Nonetheless, the skeptics of reincarnation had a "field day" and in the end, this case of past-life regression may have done more harm than good in framing the public's opinion about the factual nature of reincarnation.

However, truly scientific studies of reincarnation have been done and are continuing. Much of this work was conducted by the late Ian Stevenson and continues today with his protégé Jim Tucker at the University of Virginia.

6

The Studies of Ian Stevenson

If the Doors of Perception were Cleansed, All the World Would be Seen as it is, Infinite.

—William Blake

IAN STEVENSON (1918–2007) WAS a psychiatrist who worked for the University of Virginia School of Medicine for fifty years. He founded and directed the university's Division of Perceptual Studies, which investigates the paranormal. Stevenson's principal interest was in the arena of research into reincarnation and how ideas, emotions, memories, and physical features can be transferred from one life to another. He traveled extensively over a period of forty years, investigating three thousand cases of children around the world who told their parents they remembered past lives.

His first book on the subject was *Twenty Cases Suggestive of Reincarnation*. Half the cases were of children living in India and Sri Lanka where reincarnation is culturally accepted as fact, while the other ten cases were from Brazil, Lebanon, and from the Tlingit Indians of the southeastern coast of Alaska. A case usually started when a small child of two to four years of age began talking to his parents or siblings of a life he led in another time and place. The child usually felt a considerable pull toward the events of their previous life and they frequently pestered their parents to let them return to the community where they feel that they formerly lived. If the child made enough specific statements about their previous life, the parents (usually reluctantly) began to make inquiries about their accuracy. In some cases attempts at verification didn't occur until several years after the child began to speak of their previous life, but if some verification resulted, members of the two families sometimes visited each other and asked the child whether he recognized places, objects, and people from his supposed previous existence. In his first book, Stevenson verified that the memories of children in the study accurately recalled details from the life of a deceased person that the child had identified.

One example taken from his published reports is the case of a two-year old from Sri Lanka who overheard her mother mentioning the name of an obscure town in Sri Lanka (Kataragama) that the girl had never been to. The girl informed her mother that she had drowned there when her mentally challenged brother pushed her in the river; that she had a bald father named Herath who sold flowers in a market near the Buddhist stupa; that she lived in a house that had a skylight, dogs in the backyard that were tied up and fed meat; and that the house was next door to a big Hindu temple, outside of which people smashed coconuts on the ground. Stevenson was able to confirm that there was, indeed, a flower vendor in Kataragama who ran a stall near the Buddhist stupa whose two-year-old daughter had drowned in the river while the girl was playing with her brother who had a mental disability. The man lived in a house where the neighbors threw meat to dogs tied up in their backyard, and it was adjacent to the main Hindu temple where devotees practiced a religious ritual of smashing coconuts on the ground.

The little girl did get a few items wrong, however. For instance, the dead girl's dad was not bald (but her grandfather and uncle were) and his name was not Herath—rather that was the name of the dead girl's cousin. Otherwise, twenty-seven of the thirty idiosyncratic, verifiable statements she made were accurate. The two families never met, nor did they have any friends, coworkers, or other acquaintances in common, so if you take it all at face value, the details could not have been acquired in any obvious way.

In a second book entitled *Children Who Remember Previous Lives: A Question of Reincarnation*,[1] Stevenson describes fourteen cases typical of children who remember previous lives and details how such studies were conducted, cultural influences, and how reincarnation might explain many of the unusual abilities, inclinations, behaviors, and abnormalities in children that have no known cause.

Undoubtedly, Stevenson's greatest contribution to research on reincarnation was his two-volume, 2268-page publication *Reincarnation and Biology: A Contribution to the Etiology of Birthmarks and Birth Defects*.[2] The study contained 225 case reports of children who remembered previous lives and who had physical anomalies that matched those previous lives. Many of his subjects had unusual birthmarks and birth defects, such as finger deformities, underdeveloped ears, or being born without a foot. There were also scar-like, hypopigmented birthmarks and port-wine stains. His studies detailed cases of birthmarks and birth defects in children that corresponded to a wound (often fatal) on the deceased person whose life the child recalled that could be confirmed by autopsy findings and photos.

For example, a Turkish boy whose face was congenitally underdeveloped on the right side said he remembered the life of a man who died from a shotgun blast at point-blank range; while a Burmese girl born without her lower right leg had talked about the life of a girl run over by a train. On the back of the head of a young boy in Thailand was a small, round, puckered birthmark, and at the front was a larger, irregular birthmark, resembling the entry and exit wounds of a bullet. This was consistent with the details of the boy's statements about the life of a man who had been shot in the head from behind with a rifle. And a child in India was born with boneless stubs for fingers on his right hand. He remembered his previous life as a boy who had lost the fingers of his right hand in a fodder-chopping machine mishap.

In another case, a boy by the name of Patrick Christenson was born in 1991 in Michigan to a mother that had lost her two-year-old son Kevin to cancer twelve years earlier. Patrick's mother had a strong feeling that he was connected to her first son. She noticed that he had two unusual birthmarks that corresponded in size and shape to scars Kevin had at the time of his death. The wounds were due to medical procedures used to diagnose and treat his cancer. In addition, Patrick had a lesion in his left eye, which was the same eye that Kevin suffered blindness in due to his cancer. Finally, after Patrick began to walk he developed a limp in his left leg, which coincidentally was the same leg that Kevin had broken when he was one and a half and which caused him to limp. By the time Patrick reached the age of four he began talking about his previous life as Kevin and supplied many accurate details about his short life, and when he saw a picture of Kevin for the first time, he immediately identified the picture as himself.

In yet another case, Stevenson reported that an Indian boy named Hanumant was born with a large hypopigmented birthmark on his chest. Before his conception, his mother had seen the body of a man who had been murdered in her village by a shotgun blast to the chest. Later she had a dream that the baby she was carrying was the reincarnation of this man, and between the ages of three and five Hanumant spoke as if he was this man in his previous life. Stevenson was able to find a close similarity between the postmortem report of the shooting victim and the boy's birthmark. In Hanumant's case, the mother had a premonition that her child would be linked to this man. However, in other cases reported by Stevenson, the link between various birthmarks and birth defects was not made until after the child was born and began to talk as if he were a particular person in his previous life.

Stevenson was an expert on psychosomatic medicine and suspected that strong emotions like a traumatic death may be related to a child's retention of past-life memories. Many of the children he studied claimed that they had met a violent end in their previous life and sometimes the child's fears and abnormal behavior could be linked to the way they died. For example, a girl who claimed to have drowned in her previous life might have a fear of water, while a boy who died from a fall from a height might display a fear of heights. He came to believe that neither environment nor heredity could account for certain fears, illnesses, and special abilities, and that some form of personality or memory transfer might provide an explanation. Stevenson believed that almost everybody has experienced past lives, but only a small percentage of children retain any memories of their previous existence. Even in India, where nearly everyone believes in reincarnation, he estimated that only about one in every five hundred children have such recollections.

Stevenson was always careful to temper his scientific theories with the acknowledgment that without clear evidence of a physical process by which memories or personality traits could survive death and transfer to another body one must be careful not to commit fully to the explanation that it is due to reincarnation.

Ian Stevenson also studied other paranormal phenomena, which fall under the category of ESP. He published a review and analysis of 160 cases in which a person had a strong impression about something happening to another person who was far away. For example, a man develops a severe pain in the temple at about the same time that a relative shoots himself in the temple but before news of the event reaches him.[3]

Naturally, Stevenson had numerous critics and others were skeptical of his work simply because it did not fit their preconceived notion of how the universe works. One such skeptic was Tom Shroder, a Pulitzer Prize-winning journalist for the *Washington Post*. In the late 1990s, Stevenson allowed Shroder to accompany him on several of his trips abroad to see how he collected his data on children who remembered their previous lives. Shroder found that Stevenson conducted his interviews in a neutral and professional manner that could not implant false memories in the child or influence the parents to lie about these memories. In the end, Shroder turned from skeptic to believer and published his own book, *Old Souls: The Scientific Search for Proof of Past Lives*, in which he relates many interesting cases and his experiences and reflections from his journeys with Ian Stevenson.[4]

Toward the end of her own storied life, the physicist Doris Kuhlmann-Wilsdorf—whose groundbreaking theories on surface physics earned her the prestigious Heyn Medal from the German Society for Material Sciences—surmised that Stevenson's work had established that "the statistical probability that reincarnation does in fact occur is overwhelming. The totality of the evidence is not inferior to that for most if not all branches of science."[5] Stevenson himself was convinced that once the precise mechanisms underlying his observations were known it would bring about "a conceptual revolution that would make the Copernican revolution seem trivial in comparison." It is hard to argue with that, assuming it ever does happen.

In the West, the idea that we reincarnate is not generally accepted. As a result, Stevenson found that if children started talking about their "other family" or "other life" they were normally discouraged by their parents. In fact, Stevenson found that many of the parents he talked to were uncomfortable with the idea that their child might be the reincarnated soul of another person since the whole idea conflicted with their religious beliefs. He remarked that sometimes the children were scolded or even punished for telling their stories of a past life.

Some of the critics of Stevenson dismiss reincarnation and survival of the mind following death out of hand because they do not fit their materialist model that brain is equivalent to mind. Such critics refuse to even look at his findings, let alone to debate them in a scientific manner, because they are predisposed to reject all data suggestive of the paranormal as simply false—a product of wishful thinking. Such scientists are guilty of scientism—the improper usage of science to debunk scientific claims

7

The Work of Jim Tucker

Young children sometimes report the details of a previous life, which upon checking turn out to be accurate and which they could not have known about in any other way than reincarnation.

—Carl Sagan

DR. JIM B. TUCKER is a child psychiatrist and Professor of Psychiatry and Neurobehavioral Sciences at the University of Virginia School of Medicine. His main research interest is in children who claim to remember previous lives. He was a protégé of Ian Stevenson and after working several years with Stevenson took over leadership of the Division of Perceptual Studies when Stevenson retired in 2002. He has authored two books on children's memories of previous lives: *Life Before Life* and *Return to Life*. His books present an overview of over four decades of reincarnation research, and he has published numerous papers and appeared on several TV shows and videos talking about his work.

Like Ian Stevenson, Tucker studied cases where a child identified himself with a person having died a violent death and had birth defects consistent with wounds suffered by the deceased person. He opens his first book with the case of John McConnell, a retired New York City police officer working as a security guard when he was shot six times during an armed robbery.[1] The bullets damaged his heart, lungs, and pulmonary artery. Although McConnell was rushed to the hospital, he died of his wounds. Five years later a son by the name of William was born to a daughter of John McConnell. William bore birth defects that were very similar to the fatal wounds of McConnell—a pulmonary valve that was malformed and a right ventricle that had not formed properly because of the problem with the valve. These problems caused William to pass out on a regular basis, but fortunately, they could be corrected by surgery.

When William began talking at the age of two, he identified himself as his grandfather, John, who he had never met or known anything about. William supplied some accurate details about the life of his grandfather. For example, he knew that one of the family cats, named Boston, went by the name Boss. Tucker reports that William's mother (who would have been his daughter) felt that her son strongly reminded her of her father both in his behavior and in his interests.

His investigation of the case of Cameron Macaulay was featured in a British documentary entitled "Extraordinary People—The Boy Who Lived Before."[2] In this case, Cameron from the age of two began talking about his life as a child on the island of Barra, a tiny outpost in the East Hebrides. After a couple of years during which Cameron pestered his mother continuously about visiting the island, the show's producer heard about Cameron's claims and invited Jim Tucker, Cameron, and his mother to accompany him and a film crew to the island. There they discovered that some of the facts given by Cameron about his previous life were correct. Namely, that there was a family by the name of Robertson that owned a white house on the north side of the island, near the beach; that planes landed on the beach; and that the family had a large black and white dog. However, there was no record of a senior Robertson had been killed by a car as he was trying to cross the road as Cameron had described; plus other facts did not match Cameron's description. Tucker concluded that the case was not completely fulfilling and did not measure up to other cases he studied.

The documentary also featured a boy named Gus Taylor who lived in the United States. At the age of one and a half Gus said to his father while he was changing Gus's diaper that he had changed his diaper when he was about the same age. This incident was the start of a narrative in which Gus insisted that he was his own grandfather in his previous life—who had died a year before Gus was born. By the age of four Gus was able to identify his grandfather in an old school photo and identify his grandfather's first car from a photo. One day he mentioned that he had a sister who lived with fishes after she died at the hands of "bad guys." It turned out that his great aunt had been murdered and her body dumped in San Francisco Bay, a fact that was never discussed in the family and withheld even from Gus's father.

Another case studied by Jim Tucker that gained national attention in the news media was about a boy named Ryan Hammons who lived in Oklahoma.[3] At the age of four, Ryan began having vivid dreams about his previous life as a Hollywood agent. In stories related to his mother, he told of meeting actor Rita Hayworth, taking trips to Paris, and dancing on Broadway. In addition, Ryan provided many other details of his previous life, such as being married five times, having one daughter, and living on a street with the word "rock" in it. His mother said his stories were detailed and extensive, unlike something a child would make up. One day, when going through some old Hollywood picture books, Ryan immediately identified himself in a picture. However, the picture did not name the actor, and Ryan's mother could not find any more information about the man.

After hearing of the work of Jim Tucker, she approached him for help in identifying the man in the picture. After considerable research, Tucker was able to identify the man as Marty Martyn and confirm that he had been a bit actor turned Hollywood agent. Numerous details about Martyn's life that the boy had provided checked out. For example, Martyn had danced on Broadway, traveled to Paris, worked with Rita Hayworth, been married five times, had two sisters, one daughter, and lived on a street with the name "rock" in it (825 Roxbury Dr.). Indeed many of the details about Martyn's life and professional career as a Hollywood agent (mostly supplied by his daughter) checked out with facts supplied by the boy Ryan. In all Tucker was able to confirm fifty-five details provided by Ryan that accurately described Martyn's life. One apparent error that Ryan made was saying that he had died at age sixty-one, when his death certificate gave his age at fifty-nine. However, after examining old census records, Tucker discovered that Martyn had been born in 1903, not 1905, making him sixty-one when he died.[4]

Another interesting case studied by Tucker was that of a young boy named Hunter (not his real name) that became fascinated with golf at the age of two. Hunter received a set of plastic golf clubs at this age and began playing with them nonstop. Neither of his parents played golf but when his father inadvertently turned on the Golf Channel, Hunter immediately became more interested in watching golf than children's shows. At the age of three, Hunter identified himself as Bobby Jones in a TV show and started going by the name Bobby instead of Hunter. According to Tucker, Hunter's father showed the three-year-old a series of six photographs of famous golfers from the 1920s and asked him to identify which was Jones. When he saw the Jones picture, Hunter said, "This me," and then seeing Harry Vardon in a photo, Hunter pointed and said, "This, Harry Garden, my friend." When his father showed him pictures off the Internet of several houses including that of Bobby Jones, Hunter identified the one that Jones occupied as a youngster. Hunter continued his obsession with golf, starting lessons at the local golf course when he was still two and has become somewhat of a golf prodigy winning forty-one out of fifty junior golf tournaments by the age of seven, including twenty-one in a row.

Researchers at the University of Virginia medical school's Division of Perceptual Studies have more than 2500 reports of children from around the world who claim to remember past lives. These reports have been analyzed by Ian Stevenson, Jim Tucker, and several colleagues. The data obtained by these researchers provide convincing evidence for reincarnation. They provide solid grounds for accepting reincarnation into a worldview that rejects the materialist doctrine that mind cannot exist outside the brain. In the following chapters, we will explore additional evidence that consciousness survives death of the body.

8

They Remember their Previous Life

The soul comes from without into the human body, as into a temporary abode, and it goes out of it anew... it passes into other habitations, for the soul is immortal.

—Ralph Waldo Emerson

Besides Stevenson and Tucker, other researchers have published accounts of children that had verifiable memories of living in another body. There are also numerous accounts of adults who claim to remember their previous lives.

The Shanti Devi story

One of the first well-documented cases of a child that remembered a previous life was that of a young girl named Shanti Devi, who was born in Delhi, India in 1926. When she was four, she began talking to her parents about her previous life as Lugdi and claimed that her real home was in Mathura where her husband (Kedarnath) lived (about 145 km from Delhi). She stated that she had died ten days after having given birth to her son. When interviewed by her teacher and headmaster, she used words from the Mathura dialect and provided additional details about her home and family.

By the age of six she would not stop talking about her previous life as Lugdi. Discouraged by her parents, she ran away from home, trying to reach Mathura. Her parents were so concerned by her behavior that they took her to a doctor. The doctor concluded that she was simply a bright girl with fantasies, but he was horrified when she described with obstetric detail her difficult delivery and subsequent death. She was an only child and the doctor and her parents were mystified by how a very young girl could have known or understood such details.

A little later, her headmaster located a merchant by the name of Kedarnath in Mathura at the address she provided who had lost his wife, Lugdi Devi, nine years earlier—ten days after having given birth to a son. Kedarnath was told of the story, but even though he was Hindu, he doubted that this little girl was his reincarnated wife Lugdi. He decided to send his cousin who lived in Delhi to check out the girl's story. When the cousin arrived at Shanti's home, she happened to open the door and immediately recognized him. She then began asking about her son and supplied details about the family home and business. The cousin was so impressed that he immediately informed Kendarnath that his wife had been reborn. Intrigued, Kendarnath traveled to Delhi along with his son and pretended to be his own older brother. However, Shanti Devi immediately saw through the charade and recognized her husband and surmised that the boy accompanying him was her son. She reportedly knew many details of Kedarnath's life during the time he was with his first wife, and he was soon convinced that Shanti Devi was indeed the reincarnation of Lugdi Devi.

When Mahatma Gandhi heard about the case, he met the child and set up a commission to investigate. The commission traveled with Shanti Devi to Mathura in November 1935. There she recognized several family members, including the grandfather of Lugdi Devi. She found out that Kedarnath had neglected to keep a number of promises he had made to Lugdi Devi on her deathbed and he admitted he had found the 150 rupees she had hidden for safekeeping under the floor in a corner of their bedroom. The commission's report concluded that Shanti Devi was indeed the reincarnation of Lugdi Devi. In all Shanti made twenty-four correct statements about Lugdi's life and provided no information that was incorrect.

Also of interest is Shanti's description of her death. It reads like a classic case of a near-death experience. She described leaving her body and being in the presence of a being of light who radiated unconditional love. This was well before interest in the subject was popularized in the 1970s by Raymond Moody in his book *Life After Life*. In any case, it is unlikely that a young child could invent a story of this type that corresponds so closely with the stories of countless adults who have suffered near-death experiences.

Shot down during the war

A recent case of a child remembering his past life that gained national attention was that of James Leininger, a native of Dallas, Texas. At the age of two, James started having recurrent nightmares of being shot down during the war and unable to escape the burning wreckage of his plane—a Corsair. As memories of his previous life surfaced, he recounted many details about his former life to his parents, including the name of the aircraft carrier he flew from (Natoma Bay); his previous name (James Huston Jr.); the name of shipmates he served with, including a good friend (Jack Larsen); the name of a Japanese ship he strafed; the battle he was engaged in when he died (Iwo Jima); and the location where he crashed. James also had a mysterious knowledge of aviation, knew the cockpit layout of the Corsair, and became obsessed with airplanes. His parents checked out the facts supplied by James and found them to coincide with actual historical records. They later published a book entitled *Soul Survivor: The Reincarnation of a World War II Fighter Pilot* to tell the story of their son's recollections.

Déjà vu

Déjà vu is French for "already seen." It is the experience or feeling that one has lived through the present situation before, or the feeling of familiarity with a place that one has never visited. Reincarnation is a possible explanation for the déjà vu experience of some people. They may know their way around a locality that they are visiting for the first time. The whole place, or at least a significant part of it, may seem familiar to them. Reincarnation is not the only explanation for this common experience. A clairvoyant vision or precognitive knowledge could be involved. Of course, most psychologists today dismiss a paranormal explanation of déjà vu, instead attributing it to phenomena like "cryptomnesia," which is where information learned is forgotten but nevertheless stored in the brain, and similar occurrences invoke the memories, leading to a feeling of familiarity. Cryptomnesia is also the favored explanation of psychologists for past-life regression memories.

Past-life regression

Past-life regression is a technique that uses hypnosis to recover what practitioners believe are memories of past lives. Past-life regression is typically undertaken in pursuit of a psychotherapeutic cure—particularly of irrational fears or phobias. Although remembrance of past lives may occur spontaneously in some adults, it is much more common when undergoing hypnotic regression. As it turns out, almost anybody who can be deeply hypnotized can be regressed by the hypnotist and relate their experiences of past lives. The experiences are most often visual and take on a life of their own without intercession by the therapist. Subjects identify strongly with a particular individual and feel the emotions of that person during the regression. Often these feelings mirror problems faced by the subject in their present life. Finally, most of the subjects feel like a weight has been lifted from their mind after they relive a particularly traumatic experience from a previous life.[1]

One of the first psychotherapists to write about his work with past-life regression was Dr. Brian Weiss. He was a traditional psychiatrist who treated a patient named Catherine who had a number of chronic fears and anxieties. After eighteen months of traditional psychotherapy, Weiss decided to hypnotize Catherine in order to help her recall any childhood traumas that could be at the root of her problems. However, he was astonished, though skeptical, when she began recalling past-life traumas that seemed to hold the key to her recurring nightmares and anxiety attacks. His skepticism was gradually eroded when he was able to confirm elements of Catherine's stories through public records and was able to cure her neuroses. This was his first past-life regression and four thousand patients were to follow. Today Weiss is convinced that elements of human personality survive death and that many phobias and ailments are rooted in past-life experiences that when "relived" by the patient have a curative effect.[2]

Dr. Raymond A. Moody, Jr. has been a prominent researcher into past-life regressions. He began as a confirmed skeptic with regard to reincarnation. But as he began to investigate both psychologically healthy patients and those with phobias who under deep hypnosis suddenly began describing in vivid detail episodes from other historical periods they could not possibly have known, his interest in the meaning of these visions intensified. He found that many of his patients were cured of their phobias when they relived a traumatic event from a previous lifetime.

Questioning what his patients were describing he decided to undertake his own journey into the mysterious realm of past-life experiences. He performed self-hypnosis and relived nine former lives, including two during the Roman Empire and one as a female Chinese artist who was brutally murdered. Uncertain what to make of these seemingly genuine past-life experiences, Dr. Moody began a comprehensive two-year research project into the subject. Some of his findings are recounted in his book *Coming Back: A Psychiatrist Explores Past-life Journeys*.

Currently one of the best-known practitioners of past-life regression is Carol Bowman. She has done thousands of past-life regressions in her practice. She was first introduced to past-life regression in 1987 while living in Asheville, North Carolina. She underwent a regression to see if she could identify the cause of her chronic lung problems. In just one session, she experienced two lifetimes in which she had died due to afflictions of the lungs—dying of consumption in the nineteenth century and then in the gas chambers of World War II. That one session changed the course of her illness and convinced her that we really do live more than once.[3] She claims many of her patients' fears and anxieties are linked to past-life traumas and are helped by reliving those traumas during a past-life regression. She also claims that past-life traumas can be at the root of physical problems that can be helped by such therapy. Bowman currently oversees a popular website called the Reincarnation Forum where people can post their reincarnation and past-life experiences.[4] The Reincarnation Forum currently has over ninety thousand posts.

Bowman has also been a pioneer in reincarnation studies of children that remember their past life. Her first investigation was that of her young son, Chase, who described his own past-life death on a Civil War battlefield—an account so detailed and accurate that it was authenticated by a Civil War historian. After he relived his earlier life, Chase's chronic eczema and phobia of loud noises completely disappeared.

Inspired by Chase's dramatic healing, Bowman compiled dozens of cases and wrote her first book, *Children's Past Lives*, in which she revealed overwhelming evidence of past-life memories in children; showed how young children remember their past lives—spontaneously and naturally; and explained that such memories are far more common than most people realize. In addition she tells the reader how to distinguish between a true past-life memory and a fantasy; offers practical advice to parents on how to respond to a past-life memory; and shows how to foster the spiritual and healing benefits of these experiences.

In her second book, *Return from Heaven*, she describes her research into cases of reincarnation within the same family. She documents the emotions and relief that families experience when they discover that a deceased family member has returned in a new body as a child in the family.

Adults who remember their past lives

The vast majority of cases in which adults report memories of a previous life involve hypnotic regression. Such evidence is not very convincing since it is well known that hypnotized subjects want to please the therapist and may invent stories about a previous life in order to fulfil expectations of the therapist. Hypnotized patients are highly suggestable and simply asking them to go back to a time before their birth will usually lead them to provide a story about an earlier life. Regression therapists have noted that almost anyone who can be deeply hypnotized will recall past-life memories. For this reason and because memories of a past life can lead to anxieties—especially if the memory involves a violent death or was of an undesirable personality—the practice should only be conducted by a trained therapist with the sole intention of helping to cure that person of a neurotic fear or other mental affliction.

Memory in children is very limited and not at all continuous before the age of three or four. They lack what is termed "episodic memory," the ability to associate events in time and place. They also lack the extensive life experiences that adults can draw upon. Thus, cryptomnesia is not a likely explanation for past-life memories in children, but it could certainly explain many such memories in adults. For example, someone might recall under hypnosis (or in a trance-like state typical of self-hypnosis) many details about a previous life as a British arctic explorer by the name of Captain Sir John Franklin. They might supply extensive details about his last fateful voyage in 1845 to find a Northwest Passage, including his two ships, the HMS Erebus and the HMS Terror; the names of a number of officers that accompanied him; how they became icebound in Victoria Strait near King William Island in the Canadian Arctic; and how he and the other 128 men in his expedition died. However, all of this information might have been obtained from reading a book or seeing a documentary about the expedition that they had forgotten about.

More convincing evidence for reincarnation memories in adults comes from people that have been practicing meditation for years and perceive tidbits of memories about a previous life that make perfect sense to them because of how it fits into their current life path and interests. However, such spiritual aspirants are normally taught to ignore any information about their previous lives in order to concentrate on their current spiritual development. They are taught that we have eyes in the front of our head for a reason and that it is a blessing that we normally have no memory of a previous life.

That is not to say that some adults do not have verifiable memories of a previous life. In their book, *Past Lives: An Investigation into Reincarnation Memories*, Peter and Elizabeth Fenwick provide numerous examples of such cases.[5] In addition, Rabbi Yonassan Gershom presents compelling evidence that people living today died in the Holocaust in his books *Beyond the Ashes: Cases of Reincarnation from the Holocaust* and *From Ashes to Healing: Mystical Encounters with the Holocaust*. The evidence is based on the stories of people he counselled. He was surprised to hear powerful stories from a number of his patients of their former lives in Europe during the Holocaust. Some of Gershom's subjects provided specific and accurate descriptions of the concentration camps and death chambers for which they had no previous knowledge. In addition, Gershom tells the fascinating stories of seventy people who related to him their memories of their previous lives in his book *Jewish Tales of Reincarnation*. Other individuals such as actor Shirley MacLaine have provided vivid and detailed descriptions of their past-life experiences.[6]

Xenoglossy

Xenoglossy is the ability to speak or write a language without having learned it. Stevenson reported on a handful of cases of xenoglossy, including two where subjects under hypnosis allegedly conversed with people speaking a foreign language they did not know. If reincarnation was a viable explanation for this strange ability then one might expect many more people, including many children who remember their previous life, to be able to converse in a foreign language. This is not seen. Children's and most adults' remembrances of previous lives seem to be quite sporadic and involve a lot of imagery rather than language. Being able to recall more than a few words or phrases from a language spoken in a previous life would appear to be extremely rare. However, it might make learning that language in one's present life easier.

The medical literature has reports of people with dissociative identity disorder (DID, multiple personality disorder) where one of the personalities can speak a foreign language fluently that they had not previously studied. One such case was that of an Indian woman named Uttara who grew up in Nagpur, India speaking the Marathi language. She had an unremarkable life as a teacher at the local college until at the age of thirty-two she began to undergo personality changes and at regular intervals would take on a second personality. In this dissociated personality, she was Sharada, a Bengali woman that lived in the 1800s. When the Sharada personality took over, Uttara could speak fluent Bengali, much as it would have been spoken in the early nineteenth century, yet Uttara had only a cursory knowledge of Bengali before she began suffering from the disorder.

Although psychiatrists have proposed a number of theories to explain DID, reincarnation is not among them. However, it is possible that some of these cases, like that of Uttara's, where the dissociated personality is from the past, could be tied to reincarnation memories. Certainly, such a hypothesis is more logical than the old theory of demonic possession.[7]

9

Prodigies and Geniuses

I will return as a new and improved version, edited and redone.

—Benjamin Franklin

AS THE CHILD DEVELOPS mentally and physically it develops behaviors and character that is not solely a product of their nurturing. A good portion is inherited, but the child's karma that needs expression in this lifetime may also be a factor in their development. Hence the spontaneous appearance of extraordinary skills, ability, or genius in young children, often in contrast to the expectations of their family, may best be attributed to reincarnation.

Child prodigies

Some of the qualities that distinguish child prodigies from other children are expertise in some activity or area of knowledge; their ability to speak with emotional clarity and maturity that greatly surpasses their life experience; their extraordinary presence and awareness; their ability to focus during a conversation or on a topic without the restlessness that most children display; their exceptional memory; their ability to learn things very quickly and develop a deep understanding of a subject that would take a normal child much longer; and their ease in communicating with adults and understanding such things as adult humor. In other words, they may display knowledge and behaviors that would be consistent with an individual that was much older.

One example of an unusual ability in a child is hyperlexia. This is characterized by a child's precocious ability to read without prior training in reading—typically before the age of five. Hyperlexic children are often fascinated by letters or numbers. They are extremely good at decoding language and this aids them in becoming very early readers. Some learn to spell long words (such as encyclopedia) before they are two.

Other children seem to be born with an exceptional talent for music, math, art, language, or mechanical domains; some have extraordinary memory. Such children are sometimes given the label of savant. The so-called savant syndrome is a loose term that refers to people who have a combination of significant cognitive difficulties, often attributed to autism, and profound skills. It is believed by neuroscientists that those with this condition generally have a neurodevelopmental disorder or brain injury. One example is Stephen Wiltshire from England who was diagnosed with autism at age three and has been called a "human camera" because of his ability to draw landscapes from memory after seeing them only once. Other savants possess the uncanny skill of "calendar calculating"— quickly computing the day of the week of any arbitrary date in the past or future.

An example of a well-known savant was Kim Peek, who was the inspiration for the main character in the 1988 Oscar-winning movie *Rain Man*. Peek could read a book an hour, memorizing two pages at a time (left eye reading the left page, right eye reading the right), and afterward he was able to reproduce 98 percent of the content by heart. Still others may have a facility with foreign languages, the ability to measure distances or heights with precision without using instruments, or exceptional map-reading skills.

Still other children excel at music. For example, at the age of two, Rex Lewis-Clack's father gave him a piano keyboard. Being blind and autistic, it became his gateway to the outside world. Rex taught himself to play the piano and would play until he dropped from exhaustion. When he began formal lessons at age five, his teacher noticed his remarkable gifts. He could identify a musical note immediately, even when he heard it completely out of context. Although he cannot see and cannot read music, he only needs to hear a song once to play it back perfectly. He has whole libraries of music stored in his brain and can play them perfectly even though he might never have practiced the piece.

Neuroscientists do not have a coherent explanation for why savantism happens, but there is some evidence suggesting that savants may have experienced an undetected injury or have a defect in their brain, probably occurring *in utero* or in infancy, triggering compensatory brain development that unleashes unusual abilities. However, it is hard to explain the extraordinary ability of savants by saying they are strictly a result of brain defects. There appears to be another element involved and this can be attributed to reincarnation. Abilities learned in a previous life are carried over into the present life in such a way that a person's life path is predetermined at birth.

Jean-Louis Cardiac (1719–1726) could recite the alphabet at age three months; read Latin and translate it into French and English at four years of age; and was proficient in Greek, Hebrew, arithmetic, history, geography, and genealogy at six. Unfortunately, like a number of such children, he died very early—in his case, at the age of seven.[1] Does heredity explain such cases? No.

There have been numerous cases of child prodigies with extraordinary ability to calculate complex math equations beginning in childhood. For example, a child quickly determines using their mind that 4,294,967,297 is not prime and its divisor is 641; or a child can quickly find the product of three sets of numbers with four digits; or at age seven, already using calculus, can compute third and higher roots.

Daniel Tammet is a bestselling author and mathematical savant. One example of his amazing skill with numbers was demonstrated on March 14, 2004—international Pi Day. During the few weeks prior to the demonstration, he memorized pi (π, the unending number 3.141592...) to 22,514 places and was able to recite all the digits from memory without a single error. His extraordinary ability with numbers is called synesthesia ("sensing together" in Greek). This occurs when more than one sense such as touch and sight mix together resulting in a heightened sensory experience. Tammet is able to see and feel numbers in his mind's eye, picturing them in a three-dimensional shape with a unique color and texture. For example, he says the number fifteen is white, yellow, lumpy, and round.

The amazing abilities of calculating prodigies can be explained by differences in how their brains are "wired," but their description of simply seeing or intuiting an answer suggests that there might be more going on in the mind than can be explained by brain activity; it is not unlike how some psychics describe how they "see" things.

Sometimes a child's obsession with a sport or activity begins when they are still in diapers. This was the case for Christian Haupt's obsession with baseball. Christian's interest in the sport began when he was two and continued to grow until at the age of three he refused to wear anything besides his baseball uniform and daily begged his family to pitch, catch, and run bases with him. His parents and others observed that Christian had unusual talent for a child his age, and by his fourth birthday, Christian earned the right to be the youngest person ever to throw out the ceremonial first pitch at Dodger Stadium.

This would have been just another example of child prodigy if it were not for the fact that at the age of two he told his mother that he was once a "tall baseball player" who died because his "body stopped working." This led his mother to suspect that her son was describing the life of the famous American baseball player Lou Gehrig, who tragically died from ALS when he was just thirty-eight. Several of the past-life memories supplied by Christian were consistent with this possibility. The full story of this remarkable baseball prodigy and his memories of being Lou Gehrig are told in the recently published book (and soon to be a movie) written by his mother called *The Boy Who Knew Too Much: An Astounding True Story of a Young Boy's Past-life Memories*.

Scientists cannot currently explain why some children seem to be born with an exceptional talent, ability, or interest when there is nothing in their nurturing or genes to explain it. Is it just luck of the draw or could it be due to past-life experiences that predispose a child to such talent?

Genius

Genius is another type of an extraordinary ability that often appears to arise from nowhere. For example, Isaac Newton was not born into a family with any interest in mathematics, nor did his three siblings show any interest or ability in this area. Interestingly he was born one year after the death of Galileo and his life's work appears to be a continuation of that of his predecessor.

Mozart displayed prodigious musical ability from his earliest childhood. Already competent on keyboard and violin, he composed music from the age of five, and while still a boy performed before European royalty. Mozart received only rudimentary training in music from his father whom he surpassed in musical composition at an early age.

Most would agree that Albert Einstein was a genius, but where did his genius spring from is anyone's guess. His parents were very nurturing and encouraged him to be independent and creative, not only in science but also in music, paying for piano and violin lessons. However, neither of his parents, nor his two sisters, nor his three children showed any significant aptitude for mathematics.

A contemporary example of a budding genius might be Tristan Pang, born in New Zealand in 2001. Tristan started reading independently and doing high school math at the age of two, and was doing top-level college math by the age of nine. He started his studies at the University of Auckland and created a free online-learning platform, Tristan's Learning Hub, by the age of twelve.

Throughout history, one can find many other examples of individuals who from a very early age had a predisposition, motivation, or ability that has been considered extraordinary or a result of genius. Most of the time these extraordinary personalities arise under circumstances that defy the hypothesis of materialists that we are solely a product of our genes and upbringing. If indeed the body is only temporary but the mind (soul) is permanent, then there is no mystery why knowledge, abilities, and inclinations might be carried from previous lives to a current life.

10

The Biology of Reincarnation

To describe the overwhelming life of a tropical forest just in terms of inert biochemistry and DNA doesn't seem to give a very full picture of the world.

—Rupert Sheldrake

WHY LOOK AT BIOLOGY and evolution as it pertains to reincarnation? The reason is that the evolution of species has important elements of reincarnation within it. Few scientists today doubt that our bodies and nervous systems are the product of more than a billion years of evolution. If, as most scientists believe, the mind is purely a product of electrochemical brain activity then the only thing passed from one generation of a living organism to the next is genetic material or DNA. Can this explain how life arose in the first place from non-living chemicals? How complex biological systems and organs developed? How diverse species evolved in what appears to be quantum leaps? How complex behavior patterns or instincts are passed from one generation to another? Or how consciousness and self-awareness evolved? These questions, which represent real anomalies, are not easily explained using the physicalist approach to biology. The alternative view is that mind is nonphysical and therefore survives disintegration of the physical body and that evolution is teleological because it is driven by an organizing principle, i.e. Cosmic Mind. This approach has none of the problems of the materialist worldview and is consistent with the idea of reincarnation. Let us first review what we know about the evolution of the planet Earth and the emergence of life on its surface.

Evolution of the Earth

The story of biological evolution begins with the formation of the Earth approximately 4.5 billion years ago. Initially a hot molten body, the heavier elements such as iron and nickel settled into the core of the planet, where they are still found today in a molten state, kept warm by the radioactive decay of unstable heavy elements. The lighter substances gravitated to the surface of the planet and cooled, creating a hard crust. Meteorites and comets containing water and organic compounds including amino acids rained down on the surface of the infant Earth for hundreds of millions of years, creating oceans and continents by 4.3 billion years ago. A thick atmosphere formed consisting mainly of carbon dioxide, water vapor, sulfur compounds, methane, and nitrogen. The atmosphere helped the planet retain its liquid water and shielded potential life forms from deadly solar radiation. The conditions found on the surface of the infant Earth were conducive to the formation of life.

The first organisms that formed might have been proto-life forms that became extinct long ago, but the oldest evidence of life on earth are the fossilized remnants of cyanobacteria, dated at 3.5 billion years. These bacteria obtained energy via photosynthesis and thus utilized energy from the sun. A byproduct of photosynthesis is oxygen, and it is believed that these early single-celled organisms were responsible for converting the primordial atmosphere, which was devoid of oxygen, to one that contained oxygen. The oxidizing atmosphere was deadly to many of the microbes that were present on earth at the time, but it was a vital component needed for the evolution of more complex organisms, including animal life.

Only unicellular organisms populated the earth for a very long time (~3 billion years). Naturally, these rudimentary life forms experienced a constant struggle to survive. In this struggle, the simple unicellular organisms found it advantageous to colonize with other cells and after a long time multicellular organisms evolved. The precise time when multicellular organisms arose on the planet is difficult to determine but the best guess is that they began to populate the earth a billion years ago. This was a giant evolutionary step; not surprisingly, it took a very long time before multicellular organisms began to dominate the biosphere.

We might recall that the modern theory of biological evolution was first proposed by Darwin. He theorized that organisms evolve through the constant struggle for survival, which entails adaptation to environmental conditions and competition with other life forms. Those organisms that adapt best pass their genetic material to their offspring resulting in what is sometimes called "survival of the fittest." In this way, organisms develop greater physical and mental complexity. Life forms gradually evolve into higher and higher species, such as fungi, plants, invertebrates, vertebrates, and finally mammals.

The fossil record indicates that the path of biological evolution is not without starts and stops as many life forms come and go when they fail to adapt to changing conditions. However, the movement from less developed life forms to mentally and physically more complex life forms is predictable according to the model. Given a long enough time and fortuitous conditions self-aware life forms, which we call humans, evolved on Earth.

How long did this take? From the time of the first unicellular organisms to modern humans, the best guess is that it took four billion years. Hence the march of evolution is very slow and even our evolution from our hominid ancestors has taken approximately 250,000 years.

Problems with the materialist model of evolution

The materialist model for biological evolution may be called material Darwinism. Today this can be considered the orthodox view of evolution.

It assumes there is no need to hypothesize an "organizing principle." The evolution of life from inanimate chemicals, the evolution of organisms into complex and varied species, can be reduced to known physical and chemical processes that are passed on genetically to those organisms that adapt best to their environment. The reductionist doctrine of evolution is now accepted by nearly all biologists. One of its best spokespersons is Richard Dawkins, who wrote the book *The Blind Watchmaker: Why the Evidence of Evolution Reveals a Universe without Design.* As long as there is a credible model for evolution based on purely physio-chemical mechanisms, there is no reason to think outside the box. However, this simplistic argument does not hold up when the evolution of complex systems, structures, mind, and consciousness is considered.

One example of a complex system that cannot be reduced to individual parts is blood clotting in mammals. No less than eleven enzymes are involved in this intricate cascading system with several feedback loops. If any one of the enzymes is missing or defective, it will be a death sentence for the organism. Thus hemophiliacs, who have one defective gene resulting in the loss of a single blood-clotting enzyme, inevitably die young unless they receive modern treatments. Such an irreducibly complex system requires that essentially all the components are present before there is any advantage conferred to the organism.[1] Hence, it is difficult to explain how such a system could evolve in a step-wise manner via random mutations of DNA with selection of the fittest. The odds against all the required components of the system arising simultaneously are enormous. In addition to blood clotting, there are other systems and structures in living creatures that are considered irreducibly complex. Examples include the bacterial flagellum, the eye, and the immune system.

The materialist approach to evolution fails to account for how mind and consciousness evolved. There is no plausible physicalist theory to account for consciousness. Those in the reductionist camp explain that this does not mean they will not discover a material basis for consciousness in the future. After all, they say, it has taken science many years to discover the basis for many natural phenomena; they are working on this problem and have made some progress. However, many think the problem would seem to be insurmountable.

New York University Professor Thomas Nagel makes a strong argument that since mind and consciousness are essential features of living organisms, materialists need to explain how they evolved and why. It is not enough to say they came into existence because of physical changes in the organism. Any account of biologic evolution must explain the appearance of conscious organisms.[2] In other words, the materialist's argument that mind "emerges" after a certain level of complexity arises in a living organism sidesteps the question of where mind comes from. For mind to arise from matter implies that it exists in a potential form in matter—exactly the argument of spiritual ideology.

Another problem with material Darwinism are gaps in the fossil record. According to the theory, evolution should be slow and continuous. However, the existence of gaps in the record suggest that there has been discontinuity or unexplained "jumps" in the evolution of species. One example of this was the explosion of fauna, seemingly abrupt and from nowhere, that occurred during the early Cambrian Period. Darwin recognized that this relatively short evolutionary event (twenty to twenty-five million years) might be one of the main objections that could be made against his theory of evolution by natural selection.

Additionally, explaining complex animal instincts is problematical for the materialist model of biological evolution. While the science of genetics has made great strides in explaining how the physical attributes of living organisms are passed on between generations, genetics has yet to elucidate how complex instinctual behavior in animals is passed from one generation to the other. DNA is the genetic material responsible for the transmission of the physical structures (cells) of all living organisms (with the exception of some viruses that utilize RNA). It is the blueprint for constructing the entire organism starting from a single cell. DNA is nature's code for synthesizing proteins from constituent amino acids. Proteins are the basic building blocks of living tissues. They play a central role in biological processes. For example, enzymatic proteins catalyze chemical reactions and keep the machinery of life going. Others transport oxygen, move joints, run the immune system, and carry messages from cell to cell.

The model works for many basic animal instincts. These can be traced to how the brain and nervous system develop, and to glands and certain structures and neural networks that affect specific behaviors or cause the organism to respond to stimuli under specific circumstances. Hence genetics explains simple autonomic responses and behaviors that promote self-preservation, but how complex instinctual behavior is passed from one generation to the next is still a mystery.

Take, for example, the instinctual behavior of a bowerbird. Like the Bird of Paradise, bowerbirds have an extraordinarily complex courtship and mating dance. Males build an intricate bower decorated with sticks and brightly colored objects in an attempt to attract a mate. The bowers of all males of the species are similar, so this extraordinary behavior must be "hard wired" into the bird's brain since it is not learned behavior. In order for the behavior to be inherited, presumably it would have to be programmed into the bird's DNA. This is the same as saying that there exists a genetic form of memory. The problem is that there is no evidence directly linking DNA with memory. DNA only codes for proteins. Should we assume there is a group of special proteins that cause such complex behavior? And if there were genetic memory, then why would we not inherit knowledge from our parents, grandparents, etc.? Such knowledge is lacking and there is no known repository of inherited knowledge either in the genome of humans or in animals.

In addition, we observe complex multigenerational homing instincts in such animals as Atlantic salmon and monarch butterflies. The monarch makes a multigenerational four-thousand-mile annual trip in which descendants of the third or fourth generation know exactly where the first generation started. Females deposit their eggs during the migration and die. The larvae somehow know where their parents were heading and make the second leg of the journey. But before completing this stage they die themselves and their offspring continue the migration. Eventually after a few generations, the butterflies return to the same place where their ancestors started.

Even the relatively simple behavior of spiders in producing a web of exactly the same design is difficult to explain using inherited genetic material. In other words, the materialist explanation falls short. Today we have scientists that support the materialist model of evolution waving their hands and telling us complex unlearned behavior is passed from one generation to the next by a yet unknown mechanism involving DNA, proteins, or neural architecture. Spiritual ideology provides a simple and rational alternative hypothesis to how life first arose on earth and eventually evolved to produce human beings.

The alternative model—the roles of mind/consciousness in evolution

According to this model, it is natural for life to evolve on earth because consciousness, which is the ground substance of creation and the causal entity for mind and matter, is inherently creative. When conditions on a planet are conducive to the emergence of living organisms, then consciousness will express itself in the form of living creatures. Absent such an organizing principle, the formation of the incredibly complex combination of molecules that make up even the simplest life forms would appear to be highly improbable.[3]

Even unicellular life forms display traces of mind and consciousness. For example, a paramecium (a single-celled protozoan) can be observed in a microscope to swim about swiftly searching for food. If it bumps into an object, it recoils and darts off in another direction. Similarly, euglenas (unicellular protist) have an eyespot, a primitive organelle that is sensitive to light, and they are able to adjust their position in order to produce more food via photosynthesis. More evolved multicellular organisms show greater and more complex mental capabilities. We could conclude that even the most primitive organisms possess a rudimentary unit mind that reflects the most basic element of Cosmic Mind—mind-stuff. Previously we defined the soul as the unit mind. Hence, animals have what can be called a soul. The preexistence of this soul begins with consciousness itself—the ground substance of creation.

In undeveloped organisms, such as plants, there is only the reflection of mind-stuff and little sense of doership. As organisms develop, there is greater reflection of doership and even rudimentary expression of the "I am" quality. As "I do" feeling increases, creatures learn from their experiences and modify their behavior accordingly. For example, a dog learns a series of tricks through training. This learned behavior involves some degree of intellect, since this behavior is not instinctual for the dog. Higher animals show some degree of intellect, since they learn from experience and modify their behavior based on those experiences.

Animals such as apes, dolphins, whales, seals, etc. have a fairly well-developed intellect (intelligence). They are able to solve problems using abstract reasoning. That is, they exhibit behavior that is not trial and error but seems to involve the ability to apply previous knowledge to a new situation. Hence there is a gradual transformation of higher mental functions in animals and not a quantum leap from apes to early hominoids.

Take for example Koko (1971- 2018), a female western lowland gorilla that had an active vocabulary of more than 1,000 signs in what her instructor called "Gorilla Sign Language." This allowed her to communicate her emotions and desires with her trainer. Koko gained public attention upon a report of her having adopted a kitten as a pet and creating a name for him.

A border collie named Chaser is another example of incredible animal intelligence. Chaser learned the names of over one thousand toys. She would fetch a toy by name from a random pile of twenty or more toys. If a new toy was introduced in a pile of toys she was familiar with, she could deduce which was the unknown toy and associate that toy with its name.

But what about a bird with a vocabulary of over 150 words? A bird's brain is small and birds are not supposed to understand language—only mimic human speech. However, Alex the African grey parrot (1976-2007) was exceptional in that he appeared to have understanding of what he said. For example, when shown an object and was asked about its shape, color, or material, he could label it correctly. He could describe a key as a key no matter what its size or color and could determine how the key was different in color or size from others. Looking at a mirror, he said "what color," and learned the meaning of the word "grey" after being told he was grey. This made him the first and only non-human animal to have ever asked a question. Alex's ability to ask questions and to answer his trainer's questions with his own questions was well documented. His cognitive ability was comparable to a four-year-old child.

The "I do" capacity of mind grows in animals because of constant psychic struggles and stressful situations that force the organism to adapt and grow mentally. For example, an Australopithecus (early hominid) goes down to the river for a drink, but a crocodile is waiting submerged, hidden from sight. As she leans over for a drink of water the crocodile leaps toward her, jaws open. She barely escapes unscathed but because of this frightening experience, she learns to be more cautious when approaching the river for a drink and may pass this knowledge off to her clan members and offspring.

Once the sense of "I am" becomes predominant in a creature, it is fully self-aware and self-determinant. We call such creatures human beings. Since humans have a developed sense of self or ego, they possess freewill and can move their mind according to their desire. Plants and animals, which lack the developed sense of "I do" and "I exist," cannot act independently. Their actions are primarily instinctual. Being so guided, lower forms of life do not possess the ability to go against the natural flow of evolution. Man, however, has the ability to focus his mind in any direction he chooses, and mind always takes on the qualities of the object of its attention. This is a double-edged sword since man can choose to focus on the subtlety of Cosmic Consciousness and move forward on the path of evolution at an accelerated pace, or he can choose to direct his mind toward the crude and move backward toward the unconsciousness of animal existence.

The human being is naturally attracted to Cosmic Consciousness and ultimately will merge with Cosmic Consciousness, but from a biological standpoint, it has taken about a billion years for humans to emerge from the ocean containing simple multicellular life. How much longer will it take a human to attain divine status is anybody's guess; but just like it has taken hundreds of thousands of years for us to evolve from apes, it may not happen in a flash of biological time. The theory of biological reincarnation states that our unit mind has persisted from the time we were a unicellular organism until today, and continued evolution needs to occur before we attain unity with God.

There is no mystery why more human souls populate the earth today as compared to one thousand years ago. The hands of the clock of biological evolution turn slowly but inexorably. On this planet, we observe a plethora of creatures with developed intellect, including apes, dolphins, whales, seals, dogs, and possibly parrots. Such creatures are probably the principle source of souls that are ready to take on a human form for the first time. However, with the probability that there exist innumerable planets with advanced life in the universe one cannot dismiss the possibility that reincarnating human beings might originate on another planet in another galaxy.

While material Darwinism is the favored doctrine of most biologists today, a few biologists have challenged this doctrine. For example, Michael Behe has argued persuasively in his book *Darwin's Black Box: The Biochemical Challenge to Evolution* that the biochemistry of life is so complex and interwoven that it is incredibly unlikely that all the elements of complex organs and systems could come together merely by chance mutations when there is no advantage to an organism until a complete and working organ/system is created.

In 1981, Rupert Sheldrake published his groundbreaking book *A New Science of Life*.[4] He argued persuasively that there exists a nonlocal, nonphysical "morphogenic" field that is responsible for the form and organization of the trillions of individual cells that constitute a living organism, without which it could not function or develop from embryo to adult. In addition, biologists have studied animal behavior that appears to have some degree of cognitive interconnectedness with others of the same species. A new learned behavior is passed along to other animals that are not in direct contact with one another. This would be an example of nonlocal mental connection where learned behavior of individual animals can theoretically be passed on to others, helping to advance the evolutionary development of the species.

Summary

Reincarnation goes hand-in-hand with biological evolution. The difference between this model and the materialist model is that there is an element of our being (the unit mind) that survives death and is subtly connected to the Cosmic Mind. In other words, it is not just the DNA that is passed on from one generation to the next but also a nonphysical aspect of being, the unit mind or soul that carries a connection to the previous life. Consciousness acts as an organizing principle for all life with the ultimate goal of evolving self-aware humans that are attracted to God and after many incarnations will end their long multibillion-year journey by finally attaining union with God.

PART III

THE MIND

(AND WHY IT IS NOT BRAIN)

At the heart of the debate about whether reincarnation is a real phenomenon is the requirement that a part of our being survives death of the body. In other words, there is life after death. In addition, the unit mind must retain the karma and elements of memories, abilities, and personality from previous lives. The materialist worldview shared by most neuroscientists says that nothing of our being could survive death since mind is purely a function of brain activity and when the brain ceases to function so does the mind.

The materialist view of reality is "bottom up" ontology. Subatomic particles such as electrons, protons, and neutrons make up atoms that combine to form molecules. Complex, self-replicating biomolecules originate due to chemical transformations that give rise to simple single-celled living organisms. These simple life forms experience environmental and competitive pressures, undergo natural variations, and with increased survival of beneficial traits, evolve into increasingly complex life forms with larger brains, developed minds, and consciousness.

On the other hand, the "top-down" ontology of the spiritual worldview postulates that creation begins with consciousness and cruder aspects of reality, namely mind and matter, are epiphenomena of consciousness. In this section of the book, we will explore the overwhelming evidence that demonstrates that the spiritual view of reality is correct and that mind is not an epiphenomenon of the brain and can indeed survive destruction of the brain.

One piece of the evidence that mind cannot be equated to electrochemical events and processes occurring in the brain—cases related to reincarnation—have already been discussed. However, numerous other phenomena indicate that mind cannot be reduced to a physical basis. Some of these include a unified sense of self, memory, placebo effects, stigmata, hypnotically produced physical symptoms, mystical experience, out-of-body consciousness, near-death experiences, psychic phenomena (ESP), and elements of evolution and animal instincts. Each of these might be labeled an "anomaly" by the physicalist doctrine that mind is a product of matter since neuroscientists in this "camp" offer no reasonable explanation for them except to say they are products of a defective brain or a result of wishful thinking or even fraud.

Much of the research by neuroscientists into brain anatomy and function in the last few decades has reinforced the doctrine of a physical basis for mind. New methodologies based on behavioral, clinical, pharmacologic, genetic, neurosurgical, and electrophysiological probes as well as neuroimaging techniques, such as functional MRI, have increasingly demonstrated the close linkage between brain physiology and mental states. This has convinced most neuroscientists that almost all mental states have a physiological basis and that there is no longer any need to consider the dualistic separation of mind and brain. They also point to the fact that damage or changes in the brain cause changes in mental functioning and consciousness. However, it is not true that correlation is causation. The mind may require a healthy brain to function in the body, but this does not mean that without brain there is no mind. No doubt it has been easier for neuroscientists to study how changes in the brain affect mental states. However, the mind-brain equivalence breaks down when the study switches to how a change in a mental state (mind) affects a physical state (body and brain).

When all is said and done, however, there is a preponderance of evidence indicating that mind is not simply a product of brain and that it is a subtle, nonlocal faculty with far greater capabilities than the physical brain.

11

Unity of Self and Memories

Physicists explore levels of matter; mystics explore levels of mind. What they have in common is that both levels lie beyond ordinary sense perception.

—Fritjof Capra

The consciousness problem

CURRENTLY SCIENCE HAS NO satisfactory theory of how ordinary waking consciousness is created by the brain. Nor have neuroscientists identified any structure or group of structures in the brain that appear to generate consciousness. It is certainly true that drugs and injuries to certain areas of the brain will induce unconsciousness, but whenever there is even the faintest spark of consciousness, even in the most confabulated or dream state there will be at least some identification with self.

Clearly then, the most difficult problem facing neuroscience today is explaining a brain basis for consciousness. How and why do self-aware or sentient organisms have qualia or phenomenal experiences?

Why do we experience some things like heat and pain and other things go unfelt? Why is it that we have qualitative experiences as the brain engages in information processing from the sense organs? For example, our visual cortex is stimulated by light and we experience a beautiful pink-colored sunset; or our auditory center in the brain is stimulated by the sounds of an orchestra and we feel a blissful sensation. How can we explain how something like a mental image can evoke emotions of love, hate, or fear? How can the physical processing by the brain give rise to emotion and a rich inner life? Scientists have no good explanation of why and how experience and emotions arise in the brain, and it is not clear that they ever will. It depends on whether a brain basis for consciousness does in fact exist.

Another problem is that of intention. Intentionality depends on an "I" or user as well as a symbol for what is intended. For example, we use language to express our intention to go to the store to buy a loaf of bread. Inherent in our intentionality is the existence of an "I am" and "I do" part of our mind. The word "bread" is a symbol for the type of food we desire and is the object of our intention. The difficult problem for the "brain-equals-mind camp" is that intentionality cannot be produced by any known physical process, including the most advanced artificial intelligence (AI).

There is also the problem that in order for the sensory data that is both arriving and stored in the physical brain to be processed something has to look at it. This something could again be termed the "I" or user. This brings us to the problem of a homunculus—a little being within us who has its own system of memory and awareness along with the skills and memories needed for processing the data. This second self would need to embody all the capabilities we wish to explain in the first place.

One of the founding fathers of quantum mechanics, Max Planck had this to say about consciousness:

> I regard consciousness as fundamental. I regard matter as a derivative of consciousness. We cannot get behind consciousness. Everything we talk about, everything that we postulate as existing, requires consciousness.[1]

Planck appears to have come to this conclusion partly based on his understanding that a quantum (the smallest bit of matter or energy) exists in a state of superposition of all possible states until it is observed. In other words, observation or consciousness is necessary before anything in the quantum realm of potentiality emerges into the realm of physical reality.

The renowned physicist Henry Margenau (1901–1997) argued that the fact that different living entities all perceive the same world despite differences in their brains is evidence for the supremacy of consciousness. Since our only experience of the world is through our senses and brains, both of which differ greatly among individuals, it is remarkable that everyone perceives the same picture of the physical world. Margenau argued that this is only possible because we share the same consciousness or "One Mind" (another name for Cosmic Mind). This is the reason we perceive things the same way, and if this were not true then there would be many different perceptions of reality.[2]

Margenau may have been the first scientist to answer the question of how something nonmaterial (mind) could influence a physical object (brain). This was seen as an insoluble problem among the physicalists. If mind is nonmaterial it could not possibly affect a material object like a nerve without the expenditure of energy. Where could this energy come from? Such an exchange would violate the law of conservation of energy.

Margenau argued that the mind was nonmaterial and could function separately from the brain but was capable of influencing the brain on the quantum level. This would not require the mind to expend any energy in the process, since it is well known that the simple act of observation causes a collapse of the wave function, which could result in a particular probability event taking place in a physical organ. No energy is required, only intention or directed consciousness. Hence, he pointed out that conservation of energy as normally understood does not apply under these circumstances.

Edward F. Kelly is a noted psychologist and key author of the book *Irreducible Mind: Toward a Psychology for the 21st Century*.[3] This eight-hundred-page tome is an outstanding collection of the evidence proving that mind cannot be reduced to brain. Kelly summarizes his opinion about consciousness in the following quote from the book.

Consciousness, in short, far from being a passive epiphenomenon (of brain) seems to me to play an essential—indeed the essential role—in all of our most basic cognitive capacities.[4]

The unity of self-awareness

At the microscopic level, a person is not the same as they were even a few weeks ago. There is a continuous flow of atoms into and out of our body. Much like water flowing past a bridge, we change continuously. Thus on the most basic physical level, we are completely different from when we were a child. Yet there is something about us that is the same since we were born. That something is the sense of self-identity. Physically and mentally we have little resemblance to our childhood self, but we are aware that we are the same person at age sixty as we were at six. The continuity of self-awareness does not seem to depend on the continuity of the component parts or any physical structure.

Consciousness and a sense of self survive insults to the brain (although diminished in many cases)—even when large sections of it are removed or damaged. It is as though self-awareness results from the totality or gestalt of brain activity.

We experience unity of self-awareness, by which we mean that our experience of reality is holistic. It is not one in which each sensation remains separate and distinct from each other. This unified experience of reality is difficult to explain using the theory that brain is the basis for mind and consciousness. It is known that sensory inputs are handled by different mechanisms and/or anatomically separate parts of the brain. For example, visual stimuli, which include color, form, and motion, are handled by different parts of the brain; yet we somehow have a unified visual experience. The parts of the brain that handle sounds, touch, taste, and smell are also separate. Yet, our sense organs provide us with a unified experience of reality. We feel: "I am experiencing the outside world."

Since there is unity of experience there would either have to be what is termed "anatomical convergence" or another mechanism. Anatomical convergence means a place in the brain where everything comes together to help create the sense of self. However, such convergence does not exist in the brain and therefore the unity of self-awareness/experience must be achieved by some other means than by brain anatomy. Today the prevailing theory of neuroscientists for how separate regions of the brain are "bound" to one another involves large-scale gamma-band oscillatory electrical activity.[5] Widely separated neural populations can be synchronized by these oscillations providing physicalists with an explanation for why there is unity of conscious experience.

The problem with this theory is that there is considerable evidence that unified conscious experience continues during cardiac arrest and other stoppages of blood flow to the brain. This is observed in of out-of-body experiences and near-death experiences. Lucid consciousness with self-awareness and unity of experience is reported to continue in such cases even when there is no measurable electrophysiological brain activity. Thus the experience of being a conscious, singular person is not adequately explained by the physicalist doctrine that reduces mind to functional brain activity.

Memory

Some neuroscientists would have us believe that memory can be completely explained by electrochemical processes in the brain. However, the evidence clearly indicates that memory is a complex process that involves both the brain and the nonphysical entity we call mind. Research has shown that memories are not stored in any specific location of the brain but are spread throughout the brain much like a hologram stores a three-dimensional picture throughout its entire matrix. Memories may consist of exquisite details and mental pictures that one is not even conscious of most of the time. These detailed recollections may sometimes be brought out by electrical stimulation of the brain or hypnosis. For example, a person may recall under hypnosis a detailed description of a criminal perpetrator, including details of his tattoo, earrings, scar, and even the license plate of his get-away car, details they could not recollect following the incident. A very small minority of people have exceptional memory (hyperthymesia). Given any date, they can recall with great accuracy events that they experienced on that day.

One of the first theories for memory postulated by neuroscientists was the "trace" model. Experience causes physical changes in the brain, which can later be recalled when these physical traces or "engrams" are retraced cognitively. This model seems to work fine for simple creatures such as slugs that clearly undergo changes in their primitive nervous system in response to changes in their environment. However, such learned responses are nothing like human memory with its ability to recall at will details of past events (autobiographical memory) or general knowledge (semantic memory). In addition, neuroscientists have been unable to locate or identify any engrams in the brains of test animals. Engrams, if they exist, would be physical pathways that carry information about a past event or learned bit of knowledge. Instead, research shows that memories are not localized in a particular site in the brain. For example, in experiments using rats that were trained to run a maze, tissue was removed from their cerebral cortices before re-introducing them to the maze, to see how their memory was affected. As increasing amounts of tissue were removed memory was degraded, but remarkably, it made no difference where in the brain the tissue was removed.[6]

Another problem with this model is how memory is experienced. Most memories are not simply replays in the mind of the original event. Instead, persons will see themselves witnessing or taking part in an event. People are aware that they saw or did something. In other words, there is self-awareness, which is a mental function, not simply a replaying of events. If memories were simply stored physiologically in the brain, then they could only be replayed and not witnessed from a third-person point of view.

Along these same lines is the question of how memory is placed in time. Somehow, the mind "places" the memory in a personal time-line in such a way that it relates to other dates, names, events, etc. Thus, instead of recalling a specific memory or impression, most memories are associated with a slew of other memories and general knowledge, all of which must lie in separate physical structures according to the physicalist model. In addition, there is the question of what stamps a memory as genuine rather than imagined. It cannot be the vividness or intensity of the experience, since hallucinations may be equally or even more intense than actual experiences. People clearly possess a higher mental function that can be called personal awareness or consciousness that allows them to place memories within the greater context of experience and decide whether they are real or imagined.

A new and more viable theory of how the brain stores memories is the network model. A typical human brain weighs about three pounds but contains a trillion cells, 10 percent of which are nerve cells; hence, they number about one hundred billion. On the average, each neuron can receive signals and be connected to about five thousand other neurons. Hence, the total number of possible connections for this neuro network is ten to the millionth power.[7] This number is far greater than the number of atoms in the universe. According to the network model, memories and learning are distributed across many neurons and their connections or pathways, resulting in a network that continues to function even when parts are damaged or removed. The brain functions like a super supercomputer, and like a computer with tons of memory chips, personal memories and knowledge are "stored" in the neuro network much like files on a computer. This hypothesis is fine, but can it work without the nonphysical entity we call mind? I believe the answer to this question is no, because in order for the data stored in the physical brain to be looked at or retrieved we would need a second memory system—the dreaded homunculus that already possesses the skill and memories needed for the retrieval.

The existence of a witnessing entity for the brain or a "self" is antithetical to the brain-equals-mind theory because it implies the existence of a unitary, nonphysical mental function. Additionally, we saw earlier strong evidence for memory of past lives (noncerebral memory), which, if true, precludes the possibility that mind is brain.

With the development of modern computers, it has become popular to equate the brain with a supercomputer. Advances in artificial intelligence (AI) have made it next to impossible to determine whether one is communicating with a computer or a real person, but this is very different from saying that computers will one day attain understanding and consciousness. For example, a computer can easily translate this paragraph into Chinese by manipulating symbols and words, but it would fail miserably to provide a meaningful translation without an *understanding* of what was meant. Could it be programmed to learn the meaning of what it has translated? Critics of this computational model of mind say no; it is impossible for a machine to gain understanding and even more difficult to conceive of how it could obtain conscious awareness.[8]

12

Mind and Body

People into hard sciences, neurophysiology, often ignore a core philosophical question: What is the relationship between our unique, inner experience of conscious awareness and material substance? The answer is: We don't know, and some people are so terrified to say, I don't know.

—Raymond Moody

The effect of mind on health

IT HAS BEEN KNOWN for many years that mental states affect the body, and doctors are now taught that symptoms of disease are sometimes only in the "mind" of the patient. Psychological medical approaches, such as alternative medical treatments, a hands-on approach by a doctor or healer, and the use of a placebo may be particularly effective in bringing about a cure of such patients. Psychological feelings of hopelessness and depression are strongly correlated with an increased risk of chronic diseases such as heart disease and cancer. On the flip side, experiences of joy and laughter have been demonstrated to improve health. Studies of how mental states affect the immune system (psychoneuroimmunology) offer one possible mechanism by which one's mental state might affect one's health. There is now convincing evidence that the brain and nervous system are connected to the immune system and to the release of "stress hormones" such as cortisol. Because of the myriad connections between the brain and the body, neuroscientists today talk in terms of "mind-body unity," but what they really mean is "brain-body" unity.

There are numerous other examples of how mental states affect a physiological response or the health of an individual. For example, studies indicate that there is increased mortality following bereavement. The stress and sorrow following the death of a loved one may cause a person to give up all hope and quickly die from cardiac arrest. In some cultures, the fear from receiving a curse has similarly caused sudden death. On the other hand, there have been numerous reports of individuals that postponed their death until after a significant event such as the birth of a grandchild. Research has shown that there is a positive correlation between positive emotions and health, and that such things as meditation, imagery, biofeedback, relaxation training, and hypnosis are effective for improving disease states.[1]

The placebo (and its counterpart the nocebo) effect is fully recognized as a way in which mental expectation can affect symptoms of pain and illness. The administration of a placebo, which by definition has no actual effect on the body, creates a psychological effect, which in turn can cause a measureable physiological effect. The placebo effect is therefore opposite to the physicalist doctrine—mental states are a result of physiological changes. The placebo effect is so well established in modern medicine that it is now required to include a placebo in clinical studies designed to show the efficacy of medical treatments. Not only is the patient unaware whether they are receiving the experimental drug or treatment, in most studies the persons administering the treatment are also "blind" to whether the patient is receiving the experimental drug or a placebo. This is because if the doctor or other healthcare professional knows whether they are administering a placebo or not, they might influence the patient outcome in subtle ways. Because patients often show improvement from the placebo in "double blind" studies of this type, many experimental drugs and treatments have failed their clinical trials because there was no significant difference between the placebo and experimental groups.

One of the more curious mind-body effects is stigmata. This is where a person develops marks and sometimes bleeding wounds that Christ was thought to have suffered during his crucifixion. There have been hundreds of such cases reported in both the common and medical literature. One of the first such cases was that of St. Francis of Assisi. Persons with this affliction are most often intensely religious and have become emotionally tied to the suffering of Christ. Similar effects have also been observed in a nonreligious setting in which a person has bled from their hands, armpits, or eyes while undergoing strong emotional stress.[2] There has been no satisfactory explanation of how the brain and nervous system could produce such localized skin responses.

Another example of mind affecting physiology is false pregnancy, in which a woman believes she is pregnant but is not, and displays many of the typical signs of pregnancy. Symptoms may include abdominal enlargement, cessation of menstruation, breast changes, morning sickness, the sensation of fetal movements, and labor pains. Reports of this type of mental disturbance were more common in the past before the development of readily available pregnancy tests.

There have also been numerous reports of persons who almost overnight suffered whitening of their hair or skin in response to a severe fright or emotional stress. There is no known biological mechanism that could produce such changes, especially in hair, which except for the root is composed of nonliving tissue and is therefore not subject to physiological changes in the body.

It is well known that hypnosis in some individuals can produce surgical analgesia, changes in allergic reactions, and changes in autonomic functions such as heart rate, skin temperature, blood glucose, salivation, etc. A less known effect is on skin markings and blisters. Apparently, changes in the skin are particularly susceptible to suggestion. Conditions such as warts, eczema, psoriasis, and fish-skin disease (ichthyosis) have been cured by hypnosis.[3] Hypnosis can also induce skin conditions such as bruising, redness, blistering, and bleeding at specific sites on the body suggested by the hypnotist. Again, there is no satisfactory explanation of how such specific physiologic effects can be caused by brain-body connections.

A few exceptional individuals have been studied that can produce what is termed "skin-writing." One such person was Olga Kahl, who produced on her skin in less than a minute a communicated word or image. Her case was extensively studied and it was shown that the red color of the "writing" was well below the surface of the skin and required an exceptional control of the peripheral circulation.[4] Neuroscientists are unable to offer a biological explanation for how someone could have such exquisite control over capillary blood flow.

Studies have shown that trained meditators and yogis may have extraordinary control over otherwise autonomic processes. Examples include imperviousness to pain or cold, changes in skin temperature, and slowing or stoppage of heart and lungs. One such example taken from the medical literature is that of yogi Satyamurti. He was described as a thinly built man of about sixty years of age who volunteered to be confined to a small underground pit for a period of eight days. He claimed he would enter a state of *nirvikalpa* samadhi (nonqualified union with God) and would not need any food, water, or air during the eight days, and stipulated that no matter what happened, the pit should not be opened prematurely. The investigators fitted him with a 12-lead EKG and soon after the pit was sealed his normal heart rate began racing, eventually reaching 250 beats per minute, whereupon it stopped beating. The straight line on the EKG persisted for 7 days after which normal sinus-heart rhythm returned. Upon leaving the pit, Satyamurti's body temperature was found to be only 94.6° F. The investigators saw no tell-tail signs of an electrical disturbance, either before his heart stopped or after it restarted, which would have occurred if he had removed the leads.[5]

Transpersonal influences

One could argue that the psychophysiological effects outlined above are simply a reflection of the unity of brain and body. This explanation is far from a satisfying since it does not explain how the brain and nervous system cause such specific effects. Furthermore, this explanation breaks down completely for the many examples in which the mental state of one person influences the body of another person, sometimes without their knowledge of the event or condition. For example, a man develops a severe pain in the temple at about the same time that a relative shoots himself in the temple but before news of the event reaches him.

Sometimes the effect on another person is purely mental. Ian Stevenson published a review and analysis of 160 such cases in which a person had a strong impression about something happening to another person who was far away.[6] This would appear to be an example of extrasensory perception (ESP), but such a hypothesis would be impossible according to materialist doctrine. However, they offer no other explanation except to say that such reports are hoaxes.

Dissociative identity disorder (DID) has been studied extensively by psychologists and is of interest because of the fact that the different personalities may exhibit different physiologic conditions. For example, one personality may be allergic to a food and the other not. One may be right-handed and the other left-handed. One may have diabetes and require insulin and the other does not; or one requires glasses and the other does not.[7]

There is also considerable evidence that prayer, with or without a person's knowledge, improves medical outcomes. Larry Dossey conducted ten years of research on the relationship between prayer and healing and found compelling evidence that it can complement modern medical treatments. The scientific evidence for this is outlined in his book *Healing Words: The Power of Prayer and the Practice of Medicine*.[8]

13

Mystical Experience

I do see the Supreme Being as the veritable Reality with my very eyes! Why then should I reason? I do actually see that it is the Absolute who has become all things around us; it is he who appears as the finite soul and the phenomenal world!

—Ramakrishna

MYSTICAL EXPERIENCE IS DEFINED by mystics as the experience of union with God. Although a salient feature of the mystical experience is its ineffability, the feeble attempts to describe it use words like: clear, incredibly brilliant, timeless, unitary, ecstatic, and limitless consciousness.

Another common feature of the experience is knowledge that is nonlocal, intuitive, and not at all intellectual in nature. Mystics feel certain that they have been witness to a higher reality than that of everyday life. Such experiences are typically short-lived and may be spontaneous. They may also result from intense devotional sentiment, prayer, or meditation. People report that the experience, no matter how transient, led to a profound change in how they view and conduct their lives—a change that lasts for the rest of their lifetime.

Psychologists recognize three states of consciousness: waking or normal consciousness, subconscious or dream state, and deep, dreamless sleep. Mystics claim that they have experienced a fourth state of consciousness in which there is the experience of unity with all things. This transcendental state of awareness is said to be as different from normal waking consciousness as the dream state is from normal consciousness. The fourth or mystical state of awareness is sometimes called cosmic consciousness. What sets the mystical experience apart from other psychological states, such as hypnosis, hysteria, hallucinations, etc., is the feeling of unity with the cosmos or God. One's consciousness is not tied to the body but seems to merge or become one with the whole of creation. Even individuals with a strongly religious background describe the experience as cosmic rather than religious. They might describe feeling that they were in the presence of their chosen savior, prophet, guru, etc., but in the full-blown mystical experience, qualifications based on religious preferences disappear into the oneness of being. These common features of the mystical experience are seen despite differences in age, culture, nationality, religion, and gender. This and the fact that the experience has a profound and life-changing effect provide strong evidence that the experience is not illusory but a vision of a higher reality.

Most of the world's great religions grew from the mystical visions of such people as Shiva, Lao Tzu, Moses, Buddha, Jesus, Mohammed, and Baha'u'llah. Although the original mystical message of such prophets may have been diluted or obscured by centuries of religious doctrine and rituals, certain core mystical traditions have survived. These include the yogic and tantric practices of Hinduism, Buddhist meditative practices, Kabbalistic Judaism, Christian mysticism, and Islamic Sufism. The goal of all of these religious traditions is the same—for the individual to undergo a mystical union with Cosmic Consciousness or God. We find different terminologies used to describe this merger of the individual with the cosmic but the meaning is the same whether it is called enlightenment, self-realization, liberation, salvation, samadhi, nirvana, moksha, mukti, or satori.

A couple examples of people's feeble attempts to describe their mystical experience may help shed some light on the subject.

Robert Adams (1928-1997)

Robert Adams was born and educated in the United States. At the age of fourteen, he had his first mystical experience, which forever changed his life, and by the age of sixteen he began an earnest quest to find his spiritual teacher.

> When I had my spiritual awakening, I was fourteen years old. This body was sitting in a classroom taking a math test. And all of a sudden, I felt myself expanding. I never left my body, which proves that the body never existed to begin with. I felt the body expanding, and a brilliant light began to come out of my heart. I happened to see this light in all directions. I had peripheral vision, and this light was really my Self. It was not my body and became brighter and brighter and brighter, the light of a thousand suns. I thought I would be burnt to a crisp, but alas, I wasn't. But, this brilliant light, which I was the center and also the circumference, expanded throughout the universe, and I was able to feel the planets, the stars, the galaxies, as myself. And, this light shone so bright, yet it was beautiful, it was bliss, it was ineffable, indescribable. After a while, the light began to fade away and there was no darkness. There was just a place between light and darkness, the place beyond the light. You can call it the void, but it wasn't just a void. It was this pure awareness I always talk about. I was aware that I AM THAT I AM. I was aware of the whole universe at the same time. There was no time, there was no space, there was just I am.[1]

Gopi Krishna (1903-1984)

Gopi Krishna was a yogi mystic born in India who began practicing meditation at the age of seventeen. He wrote about his mystical experiences, which he attributed to the rising of kundalini energy, in his autobiography, *Living with Kundalini*, beginning with his first such experience.[2]

> Suddenly, with a roar like that of a waterfall, I felt a stream of liquid light entering my brain through the spinal cord. Entirely unprepared for such a development, I was completely taken by surprise; but regaining self-control instantaneously, I remained sitting in the same posture, keeping my mind on the point of concentration. The illumination grew brighter and brighter, the roaring louder, I experienced a rocking sensation and then felt myself slipping out of my body, entirely enveloped in a halo of light. It is impossible to describe the experience accurately. I felt the point of consciousness that was myself growing wider surrounded by waves of light. It grew wider and wider, spreading outward while the body, normally the immediate object of its perception, appeared to have receded into the distance until I became entirely unconscious of it. I was now all consciousness without any outline, without any idea of corporeal appendage, without any feeling or sensation coming from the senses, immersed in a sea of light simultaneously conscious and aware at every point, spread out, as it were, in all directions without any barrier or material obstruction. I was no longer myself, or to be more accurate, no longer as I knew myself to be, a small point of awareness confined to a body, but instead was a vast circle of consciousness in which the body was but a point, bathed in light and in a state of exultation and happiness impossible to describe.

Gopi Krishna goes on to describe how the serpentine energy (kundalini) transformed his life, creating many difficulties and wonders.

In religious scripture, the most concise description of the mystical state might be that of the Mandukya Upanishad. It says that below the waking state lie the states of dreaming sleep and deep dreamless sleep. But beyond these is the fourth state (*turiya*), the transcendental state, which is characterized by pure unitary consciousness and bliss (ananda)—invisible, otherworldly, incomprehensible, without qualities, indescribable, the unified soul in essence, peaceful, auspicious, and without duality.

Mystics describe their experience as entering a "timeless state" or "eternal now." Apparently, during a mystical experience the flow of time ceases. Many also claim to experience a feeling that the entire universe lies within their being. There is a realization that the entire cosmos is the thought projection of the Godhead. For what may last only a moment, time seems to stop and the boundaries of their being dissolve into the One—there is no longer any difference between them and God. Universally they describe coming to the realization that our perception of reality is illusory—a product of perceiving a relative reality. Ultimate reality is both limitless and timeless. Interestingly, their perception of reality goes hand in hand with the modern concept of reality envisioned by Albert Einstein.

The spiritual implications of Einstein's theory of relativity

In 1887 two American scientists, Albert Michelson and Edward Morley, showed that the speed of light was a constant whether the Earth was moving toward or away from a distant star. This observation contradicted the commonsense notion that speeds should add up—e.g. a bullet fired forward from a fast-moving car should have a higher velocity than one fired backward. For light, this is not the case—it travels at a constant speed in empty space.

Albert Einstein realized that if the speed of light was a constant no matter what point of reference was used, then something else had to change to account for its constancy. He sensed that this "something" must be space itself. He proposed that space could flex and change, become compressed or expanded according to the relative motion of an object and an observer. The only constant was the speed of light itself or an integrated four-dimensional "fabric" he called "space-time." These insights led to Einstein's general theory of relativity, which states that the universe has four dimensions. There are three of space—width, length, and height—and one of time. Time is not a separate dimension in this scheme but is fully integrated with the three spatial dimensions. In other words, each of the four dimensions of space-time has a spatial and temporal component.

When Einstein first published his paper on general relativity in 1907, saying that with motion, space shrinks and time slows and that space could be bent by a massive object, he was not taken seriously by most of his peers. However, in May 1919 a team led by the British astronomer Arthur Stanley Eddington photographed stars close to the Sun during a solar eclipse on Principe Island located near the Equator in the Atlantic Ocean. Eddington found that starlight from stars close to the Sun was bent by the Sun just as Einstein predicted. This astounding confirmation of his theory guaranteed Einstein's global renown.

Einstein's general theory of relativity describes gravity as a geometric property of space-time caused by objects with mass. Gravity is seen as nothing more than a distortion of space-time. The more massive the object the more it distorts or curves space. Such curvature of space caused by a massive object such as a star causes light passing near it to bend, and if the mass is great enough the gravitational force and resultant bending of space will be so great that nothing can escape its pull, including light—creating a black hole.

Einstein's equations indicated that the faster an object moves, the slower the passage of time and the more mass it gains. Ultimately, at the speed of light, time stops. However, for matter it would be impossible to attain this speed since it would require all the mass-energy of the universe. Photons, however, which carry electromagnetic radiation such as visible light, can move at the speed of light since they have no mass. Their internal clocks are stopped and they do not decay like other particles.

Experiments have proven Einstein's theories about space, time, energy, and mass to be correct. Astronomers have identified thousands of black holes, and physicists have shown that the rate of decay of unstable subatomic particles accelerated near the speed of light in a cyclotron is slowed and the particles gain the exact amount of mass as predicted by the theory.

Einstein's theory of relativity with its mathematical description of a four-dimensional space-time continuum has been verified by numerous other experimental observations. Furthermore, predictions made by the theory have proven correct and highly accurate. The most recent example is the confirmation of gravitational waves that Einstein predicted. The miniscule distortion of space-time that was produced by the collision of two black holes in a distant galaxy was picked up by the pair of LIGO observatories in Hanford, WA and Livingston, LA in 2016.

Einstein's theories stand as one of the most important scientific discoveries of all times, and many of today's technological advances (e.g., the GPS system) depend on the relative mechanics derived from the theory. Several startling and bizarre consequences arise from this new model of the universe. Below is a summary of some of these.

- *Within integrated space-time only events have meaning.* An event such as the explosion of a supernova in our neighboring Andromeda Galaxy is considered a point in the four-dimensional matrix of space-time. This event may be observed on Earth 2.5 million years later—the time light would take to travel the distance. However, for the stream of light particles (photons) carrying the information of the event no time will have passed during their transit because their clock is stopped. From the perspective of the photon, distance does not equate to the passage of time, while for us it does because we are accustomed to equating time with distance. Within four-dimensional space-time, the exploding of the star is a singular event.
- *Movement causes space to convert to time.* When an object is not moving relative to another object, then it is moving in time alone. If an object is moving near the speed of light, then it is moving mostly through space and its clock will slow down relative to a stationary clock. For example, in the future, human beings might develop a spaceship that can travel at 90 percent of light speed, which is 270,000 km/sec. After the astronauts reach full speed on their way to a planet circling a star twenty light years away, they will calculate the distance to the star to be only ten light years (because of the compression of space). They will then calculate that it should take eleven years to reach

their destination. However, because their clock runs at half the speed as clocks on Earth, the event of their arrival after eleven years of their time will correspond precisely with their expected arrival by earthlings (twenty-two years).

- *A massive object distorts space-time.* Like a bowling ball bending a rubber membrane, the bending of space-time corresponds to gravity, and the distortion pulls on time as well. Time is slowed down near a massive object. This effect was demonstrated scientifically by synchronizing two atomic clocks and moving one to the top of a tall building for a week. Upon return to ground level, the upper clock was found to have run a little faster than the one on the ground because the force of gravity diminishes with distance above the surface of the earth. If astronauts orbited a black hole with its massive gravitational pull and then were able to return to Earth, their clock would run significantly slower during their close approach compared to clocks on Earth. It is conceivable that after a few hours of "slowed" time near the black hole they could return to Earth and be younger than their grandchild is.
- *Spatial dimensions are compressed at high speed.* The shape of an object such as a spaceship traveling near the speed of light would look compressed or flattened to someone observing it as it passed by Earth. To the astronauts on the spaceship, everything would look perfectly normal since everything including their measuring devices would have shrunk the same relative amount.
- *Space and time are observer dependent.* Time and length may expand or shrink depending on the relative state of motion of the observer and the observed. As space shrinks, time expands (slows). Space is transformed into time and time into space. This is the hallmark of a four-dimensional substance in which the dimensions have both spatial and temporal aspects that are fully integrated and inseparable.
- *The "now" is not the same for observers moving relative to one another.* For example, if astronauts were traveling away from Earth at high speed their experience of now would be of events that already occurred on Earth, while if they were moving toward

Earth they would experience events that have not yet taken place on Earth. The now experienced by a moving observer is just as valid as that of a stationary observer. Hence, the now, just like the past and future, is observer dependent and mutable.
- *Four-dimensional space-time is unchanging and characterized by wholeness.* From the three-dimensional perspective of human experience everything changes in time, but underneath this relative reality lies the unchanging four-dimensional reality of space-time.

It is this last and most mindboggling quality of reality suggested by the theory that is applicable to the question of whether our perception of the flow of time is illusory. The theory of relativity implies that beneath this ever-changing realm of human experience lies a deeper, singular, and timeless reality. Scientists call this new picture of space and time "block time." Events, whether past, present, or future are at fixed points in this four-dimensional reality. For us time is associated with movement. However, just as the rapid movement of still pictures on a cinema screen creates the illusion of movement, similarly we experience the flow of time as our three-dimensional world moves at a constant rate through the fixed four-dimensional matrix of space-time. This model suggests that "cosmic events" are predetermined but it does not require that for individuals their fortunes be "set in stone" since as humans we can exercise our free will.

If relativity theory is correct then the human mind might be able under the right conditions to transcend ordinary consciousness and attain "four-dimensional sight." This perception might be described as a "higher reality," both timeless and limitless, hallmarks used by mystics to describe their experience.

Summary

There have been hundreds of accounts of mystical union described in popular literature. The experiencers universally say they entered into an indescribable state of conscious awareness that was unitary, ecstatic, and timeless. In this state, they report that their little self disappeared as they merged with God. Their descriptions of this life-changing event are strikingly similar. Attempts to label their experience as hallucinations or products of abnormal brain chemistry would have to come to terms with this fact, and the fact that the experiences are both incredibly powerful and transformative. Mystics claim to have experienced a state of consciousness that is more expansive, knowing, and unitary than normal waking consciousness, and they are seemingly witness to the supremacy of consciousness over mind and matter. They describe Cosmic Consciousness as the foundation for reality and creation as an ongoing unfolding of Consciousness into the material world. Their description is contrary to the physicalists' doctrine that mind and consciousness are byproducts of matter (brain). Unfortunately, the experiences of mystics are personal and cannot be shared directly with us so there will always be skeptics who will label such experiences as "anecdotal" or products of a deranged or abnormal mind.

General relativity has become an integral part of our understanding of how the universe formed and how it evolved from a dimensionless point. The philosophical implications of the theory might explain important aspects of the mystical experience, but few scientists today want to talk about these implications, probably because to do so would be to admit that our experience of a three-dimensional world is merely a shadow of a far richer, more encompassing, timeless realm. This is the same picture of reality painted by idealist philosophers such as Plato and would be contrary to the doctrine of materialism that is popular in the scientific community today.

14

Out-of-Body and Near-Death Experiences

The universe begins to look more like a great thought than a great machine.

—James Jeans

What is an OBE?

During normal or waking consciousness, we have a clear sense that we are "in" our body. This awareness is altered in what is called an out-of-body experience (OBE). Here a person has a vivid experience of leaving their physical body. Often they feel themselves floating above their body and are able to witness events from this unusual perspective. The OBE is a common feature of the near-death experience, but differs in that the OBE is usually not associated with a close brush with death. OBEs are most likely to occur during anesthesia, while falling asleep, and during lucid dreams. Psychologists label the

OBE a dissociative experience arising from different abnormal psychological and neurological factors—an altered state of consciousness like a dream or hallucination. Few neuroscientists today consider the OBE to be evidence for the ability of the mind to leave the body and gain information that would not be available locally through the sense organs. Investigators also report that many of the features of an OBE can be reproduced using artificial stimulation of the brain and by hallucinogenic drugs such as LSD. The consensus among neuroscientists is that OBEs are of physiologic origin. But is this correct? While some of these experiences may have a psychophysiological explanation, there are good reasons to believe that a paranormal explanation is needed for some OBEs.

For example, let us consider what happens in a typical OBE as reported in the medical and popular literature:

> A young woman is admitted to the hospital suffering from acute appendicitis. While under anesthesia for this routine operation to remove her appendix, she suddenly experiences ventricular fibrillation. The hospital staff rush to try to restore her heart to normal sinus rhythm since this condition halts all blood circulation to her brain and she will certainly die if left untreated. The staff manages to restore normal rhythm to her heart using a defibrillator, and complete the operation without further complications. Following the procedure, she tells the medical staff and members of her family that she experienced a lucid sensation of floating above her body, and from that vantage she could clearly hear and see the things that occurred as the doctors and nurses frantically tried to restart her heart. She accurately describes the color of the mask (red) and the resuscitation bag (blue) that was used over her mouth; the color of the paddles (orange) the doctor used to shock her heart; the color of the hair cover (light purple) on her head; the layout of the operating room, scribbles on the surgery schedule board, the nurse's hairstyle, and even the fact that her anesthesiologist was wearing mismatched socks. Her perceptions turn out to be remarkably accurate despite the fact she was totally blind.

There have been literally thousands of people who have reported observing events from a perspective outside their body. H. Hart did an analysis of almost three hundred published accounts of OBEs in which the person reported observing events that they could not have witnessed using their ordinary senses, and the details in nearly one hundred of these cases were corroborated by a second party or by other means.[1] Hart also investigated OBEs in which a second person at a distant location had a vision of or felt the presence of the person undergoing the OBE. Such cases are called "reciprocal apparitions." If the knowledge received by persons undergoing an OBE were strictly due to an ESP phenomenon known as remote viewing (a form of clairvoyance), then it would be difficult to explain how another person could share in the experience.

Beginning in 2001, Sam Parnia and colleagues have investigated claims of OBEs by persons undergoing cardiac arrest. These so-called AWARE studies (AWAreness during REsuscitation) have looked into the experiences of over 1500 cardiac-arrest survivors in order to determine whether people without a heartbeat or brain activity can have documentable OBEs and near-death experiences. In some of these studies, objects or placards were placed on shelves that could only be observed from above. So far, only about 2 percent of the individuals in the study have reported full awareness compatible with an OBE and none have identified the hidden image. However, most of the OBE patients described having clear visual experiences from a vantage point outside their body, something Parnia claims would be impossible when the brain is shut down during cardiac arrest.[2] It seems likely that if the OBE is real, eventually someone will report seeing the hidden image providing further evidence supporting the reality of this experience.

For adepts in yoga, the out-of-body experience is taken for granted as one of the siddhis (powers) that may be acquired through intensive meditation. The term used to describe this supernormal power is "astral projection." It is believed that the subtle or psychic body leaves the physical body but is still tied to the physical body by a subtle cord—much like a kite is attached to the ground by a string. The psychic body is then free to roam the cosmos according to the will of the yogi.[3] Similar to other powers obtained by spiritual practice, students of yoga are constantly reminded not to pursue them because they can lead one astray and be dangerous to their psycho-spiritual health. Today several institutions specialize in the study of OBEs and astral travel.[4]

Near-death experiences

The near-death experience or NDE is characterized by a lucid out-of-body experience along with feelings of peace and bliss. Similar to the OBE, there will usually be accurate visual and auditory experiences from a vantage point outside the body, often seeing their lifeless body from above and witnessing attempts to resuscitate it. Next, there is usually the experience of entering a dark tunnel with a brilliant light beyond. Often loved ones are seen in the tunnel. Once they enter the light, there is the feeling of being in the presence of a being that radiates infinite, eternal, and unconditional love. Often these experiences are complimented by a life review in which the dying person witnesses or relives thousands of past events simultaneously in an altered state of consciousness best described as an omni-view of life events. Often the events are not what they consider as the most important ones in their life, but trivial events where they see how their actions affected other people both positively and negatively. Finally, the person may feel that there is a line that if crossed will lead to death. They either feel or are told that it is not their time to go and describe being reluctantly drawn back to their body—the usual reason being commitment to family or loved ones. Almost universally, the person will describe the NDE as a life-changing event. Their attitudes, beliefs, and outlook on life are permanently and dramatically changed. Even among persons who were previously atheists, there is a certainty that God exists and that there is life after death.

NDEs are normally associated with a close brush with death, such as cardiac arrest, but may also occur during anesthesia or be triggered by a strong expectation that they are going to die but in which there is actually no physical possibility of death. During the NDE, they claim to experience being in a nonphysical body that is completely healthy, pain free, weightless, and blissful. They report being fully conscious and have full memory, judgment, and imagination. The images they witness in the "disembodied" state are described as highly vivid, with lucid awareness and clarity—often described as more real than normal waking consciousness.

The history of NDEs goes back thousands of years and they have been reported in many diverse cultures. One of the earliest accounts as mentioned earlier was reported by Plato. Paul's experience on the road to Damascus sounds like it could have been a NDE. In modern times, numerous such reports have surfaced both in the popular and medical literature. Estimates place the prevalence of NDEs at 10–20% of patients close to death.[5]

Raymond A. Moody, Jr. has probably done more to popularize the NDE than any other researcher. In his bestselling book *Life after Life*, he chronicled the accounts of 150 survivors of a near-death experience and first coined the term NDE to describe these accounts. Moody concluded that the NDE was strong evidence for life after death.

More recently, Jeffrey Long in his book *Evidence of the Afterlife: The Science of Near-Death Experiences* argued that there are nine observations that prove the existence of life after death.[6] These were generated through his study of hundreds of NDEs and the consistencies of the accounts that he compiled over the years. These arguments include that it cannot be medically explained how people experience consciousness outside their body when they are clinically dead; secondly, blind people experience visual perceptions during their NDE; thirdly, children give details of their NDEs similar to adults, though they may have never been exposed to this concept; and finally, "life-review" experiences that reflect real events. These observations are the primary basis for Long's assertion that the NDE data proves that there is life after death.

One of the most compelling reasons for believing that the NDE is a real phenomenon associated with the dying process is the remarkable similarity of the accounts, regardless of age, nationality, religion, race, culture, and other demographics. No two experiences appear identical, but even among children one or more of the elements mentioned above are reported. Certainly, predisposed cultural and religious attitudes play a role in the experience, since for example, Christians are more likely to say that they felt or saw an image of Jesus while Hindus are more likely to report seeing Krishna.

The NDE clearly challenges the prevailing opinion of neuroscientists that all mental activity can be attributed to the brain. What is most difficult to explain is how there can be a continuation of consciousness and even enhancement of mental awareness during a time when the brain is shutting down or which in many cases has stopped functioning because of a stoppage of blood flow. Not only are there lucid consciousness and vivid memory but also fully structured thought processes and the same sense of self that exists in normal waking consciousness. Often the experience is described as so beautiful and transcendent in nature that words simply cannot describe it, and its effect on the person may be felt for decades.

EEG studies of people suffering cardiac arrest indicate the absence of gamma waves, normally associated with waking consciousness, and within a few seconds of circulatory collapse the EEG will display a flat line, which is one of the characteristics of death—the others being lack of heart beat, respiration, and brainstem reflexes. As mentioned before, Sam Parnia, a cardiologist who has done extensive research on the resuscitation of patients suffering cardiac arrest, has concluded that the oxygen-deprived brain of such a person could not possibly produce the images and lucid consciousness that is a hallmark of the NDE. In fact, the ordinary unconsciousness that accompanies such events is typically associated with confusion and impairment of memory (amnesia). It is also very difficult to explain how persons undergoing such insults to the brain could obtain and later relate accurate and verifiable information about events that took place while they were unconscious and in some cases considered clinically dead.

Physicalists' explanations

The OBE and NDE are a direct challenge to those scientists that believe that mind is generated by brain activity. In order to feel at ease with their preconceived belief that mind cannot possibly be nonmaterial, since nature does not work that way, they had to come up with several alternative explanations for the NDE. These are outlined below.

- *During the process of dying, as brain shuts down, people may experience a dream-like state and/or hallucinations as the subconscious mind takes over.* However, this theory fails to adequately explain how people who are pronounced clinically dead could have such lucid consciousness, a consciousness that is described as more vivid than normal day-to-day consciousness—nor does it explain many of the other elements of the full-blown NDE.
- *A flat-line EEG does not guarantee lack of brain function.* There is ample evidence to suggest that there can be deep-seated neuronal activity in the brain that does not show up on a scalp EEG. Hence undetected brain activity could be going on and be responsible for the NDE. The problem with this theory is not that there might be brain activity, but whether there could be the type of brain activity that is considered necessary for conscious experience. The answer appears to be no, since the hypothesis that there is a brain basis for consciousness depends on the ability of widely separate regions of the brain to coordinate with one another—something that could not occur under these circumstances.[7]
- *The experiences people report during an OBE or NDE do not occur when they seem to occur.* Either they could occur during the initial shutting down of the brain due to a lack of oxygen or right afterward as the brain is being restored to normal functioning. One problem with this hypothesis is that such insults to the brain are known to be associated with confusion and memory loss. Secondly, many NDEs have "time stamps" in which the person reports specific events during the time their heart was stopped. These include witnessing resuscitation attempts and frantic attempts by others to call for help.
- *Many of the features of an OBE or NDE can be reproduced by drugs or other means.* For example, the massive release of endorphins and enkephalins (endogenous opioids) in the brain might cause blissful feelings and an altered state of consciousness.

The administration of ketamine (an anesthetic) is capable of reproducing many of the features of the NDE. High G-forces, such as those induced in a centrifuge for flight training produce similar sensations to a NDE, including tunnel vision, lights, and floating sensations as consciousness is slowly lost.
- *There are obvious similarities between OBEs and NDEs and hallucinations caused by psychedelic drugs such as LSD and DMT.*
- *The OBE and NDE are not paranormal phenomenon.* Every element of these experiences can be replicated with drugs, with anoxia (lack of oxygen), with lack of blood flow, or by turning off circuits in the brain.

The biggest challenges facing such skeptics is how any of these proposed physiological explanations can produce experiences that seem absolutely real, memorable, provide accurate images of events that took place while unconscious, have lasting and profound aftereffects, and produce a full spectrum of NDE experiences. None of the proposed physiological explanations—hallucinations, dreams, confabulations, drug-induced brain effects come close to producing these effects.

Another challenge to the brain-based explanation of the NDE comes from Raymond Moody. He points out that elements that we think of as a near-death experience—leaving one's body, going into the light, seeing a panoramic review of one's life, and seeing deceased relatives—occur not just to people who have the NDE and are brought back, but to healthy and uninjured bystanders at the bedside of the victim. These so-called "shared death experiences" are not uncommon, and the bystanders say that they also left their bodies and accompanied their dying loved ones partway toward the light, or that they saw the apparitions of relatives and friends of the dying person. Moody has accounts of hundreds of people with shared death experiences—some of whom had the identical experience to the person with a NDE. They reported experiences characteristic of the NDE but they themselves were perfectly healthy.[8] If the cause of these near-death experiences were a physiological insult to the brain, how could bystanders who were not ill or injured have essentially the identical experience?

Summary

The most popular current hypothesis of neuroscience is called epiphenomenalism, which in effect says that consciousness is dependent on the brain and the electrochemical events going on in the brain. Hence, most neuroscientists freely dismiss the OBE and NDE as aberrant brain conditions. Neuroscience has yet to propose a meaningful model for how consciousness and self-awareness can arise from strictly neurochemical reactions in the brain. However, it would not be fair to underestimate the great progress made by neuroscience in the last few decades in the understanding of the brain and how it functions, and the indispensable role it plays in the perception of reality. Researchers in the field of neurophysiology have uncovered strong evidence that almost all mental states and conditions are accompanied by changes in the brain. However, just because mind is dependent on brain for its function does not mean that brain is the cause of mind.

If one accepts the prevailing doctrine that matter is the fundamental building block of reality, then it is natural to overlook the distinction between cause and effect and dismiss all the evidence to the contrary as anecdotal, superstition, bad science, or simply impossible. The history of science has many examples in which the large majority of scientists favored the status quo even in the face of anomalous evidence that in hindsight should have offered clear indications that the prevailing theory was incorrect.

The data on OBEs and NDEs indicates that these experiences involve altered states of awareness and vivid and unforgettable images and sensations—mental functions that could not possibly occur according to current neurophysiologic models of how the brain produces mind. The alternative hypothesis that mind is not body and may separate from it under certain conditions, including death, appears to be a more plausible explanation for these experiences.

15

ESP

There is no reason why an extraphysical general principle is necessarily to be avoided, since such principles could conceivably serve as useful working hypotheses. For the history of scientific research is full of examples in which it was very fruitful indeed to assume that certain objects or elements might be real, long before any procedures were known which would permit them to be observed directly.

— David Bohm[1]

ESP IS SHORT FOR "extrasensory perception" and consists of several related phenomena. These are clairvoyance, telepathy, precognition, and psychokinesis. Clairvoyance (remote viewing) is the ability to obtain information about places, things, or events at a remote location. Telepathy is the transfer of thoughts or feelings between individuals at a distance without any apparent physical means. Precognition involves perceiving information about future events before they occur, and psychokinesis is the ability of the mind to influence matter, space-time, or energy. These phenomena are also known as psi phenomena and research into them is known as the field of parapsychology.

Telepathy

The word telepathy actually means "feeling at a distance." However, since it was first coined by renowned psychologist F. W. H. Myers in 1882, it has come to mean communication between two minds. Experimental studies of telepathy have been going on for at least 130 years and it is safe to say that if no positive data were obtained studies of this type would have been discontinued long ago.

Probably the best example of the modern research method for telepathy is the ganzfeld method, which means "complete field" in German. It is well established that a quiet, peaceful environment with limited sensory inputs is optimal for receiving information from another mind. Hence the receiver in such experiments is situated in a comfortable reclining chair with halved ping-pong balls placed over their eyes and white noise playing on headphones. A soft red light illuminates their face. These conditions create a restful state similar to one produced in a sensory deprivation chamber or floatation tank. Next, an assistant is asked to select one target pack of four images from a pool of many such packets. Then one target image is selected from the four images in this packet, which is in an opaque envelope. This envelope is given to the sender, who is in a distant, isolated room called the "sending chamber." The sender unseals the envelope and mentally tries to send the image to the receiver over the next fifteen to twenty minutes, after which the receiver is returned to a normal environment and asked to select the image that was "sent" from the four images that were in the original packet. Variations to this procedure include using video clips and automating the selection process using a computer. The sender is asked to rank the images from one to four. If the actual image (or video clip) is ranked first, it is considered a hit, which is expected to occur by chance one fourth of the time.

From 1974 to 1997, there were over 2500 ganzfeld sessions reported in forty publications by researchers around the world. In a 1985 meta-analysis[2] of the published studies that provided hit data, twenty-three out of twenty-eight studies resulted in hit rates greater than chance with odds against chance of a billion to one.[3] After receiving some criticism from skeptics, a new round of experiments were started in 1983 by Honorton and colleagues that were computer-controlled. During the six years of the study using 354 volunteers, a 34 percent hit rate was achieved.[4] These results were similar to the nonautomated experiments indicating that people can sometimes receive information at a distance without the use of the five senses.

There have been numerous, well-documented reports of telepathy between identical twins—some separated by birth and later reunited. The only reasonable explanation for the uncanny connection that clearly exists between some twins is that their minds are entangled, just as their bodies were in the womb.[5]

Clairvoyance

The difference between clairvoyance and telepathy is that no sender is required. Information is obtained about something from a distance, and thus clairvoyance is also called "remote viewing." There have been many anecdotal reports of clairvoyance involving a dream, sudden knowledge, or a strong feeling that a loved one was injured or had died. The time and nature of the perceived incident often coincides with details of the accident, well before they receive news of the event.

The first scientific studies of clairvoyance began in the 1880s using targets such as playing cards. Over the next 130 years, the experimental techniques have been refined in order to better control the experiments, and five-symbol card decks have become standard. The cards may be in sealed envelopes, behind opaque screens, separated by distance, or even displaced in time. The hit rate using all these methods in a subset of tightly controlled card experiments run by twenty-four different investigators from 1934 to 1939 in over nine hundred thousand trials was significantly above the 20 percent expected by chance.[6] J. G. Pratt calculated that if all the clairvoyance card tests conducted from 1882 to 1939 were combined, the odds against chance would be more than a billion trillion to one. This was a four-million-trial database performed by sixty investigators in many different countries and reported in nearly two hundred published reports.[7]

Precognition

There are innumerable examples of people who have accurately prophesized their own imminent death, a disaster, or other future events. Various cultures and religions take prophesy quite seriously and it forms the basis for many mythologies. Even today, while it is common to dismiss many of the stories of lore as myths and superstitions, it is such a common experience for people to have an intuitive feeling about something that later happens, that prophesy and fortune telling are accepted by a majority of people worldwide. This does not mean precognition is true, but it does suggest that it may not be that rare for a person's mind to bump into their future.

However, the experimental "proof" of precognition comes from thousands of well-controlled scientific experiments. The typical experimental protocol is to have a person guess in advance which target will be displayed from a randomly selected group of possible targets. If the correct target is selected, then this is counted as a hit. Many studies of this type have been conducted since 1935, and in almost all of the studies, the null hypothesis (that results are due to chance) was rejected. Combining all the results from 309 studies, 113 articles, sixty-two different investigators, and almost two million individual trials gave odds against chance of 10^{25} to one.[8] Odds of even ten thousand to one indicate that chance occurrence must be eliminated as an explanation for the results.

Experiments also show that unconscious autonomic responses to stress occur before the stressful situation is witnessed consciously. The technical term for such a response is "presentiment." It can be measured by attaching sensors such as skin conductance, heart rate, or blood flow in the fingertip, and then having a computer display a series of photos, some of which are calming and some of which are disturbing. Naturally, the monitors show a strong response immediately after the subject sees a disturbing photo and no reaction following a calm photo, such as a field of flowers. The surprising thing is that the computer registers a significantly larger fright-and-flight response just before the emotionally charged photo is shown than before the calm photo. The opposite effect of a dropping heart rate can also be seen just before the calm photos are shown.[9] Experiments like this in which an unconscious autonomic response precedes the actual insult have been replicated and indicate that the nervous system is capable of anticipating future events.

Psychokinesis

Is there such a thing as mind over matter? In a sense, the answer is easily demonstrated. Think about raising your arm and it is raised. If we assume mind is nonmaterial (not brain-based) then it moves matter (body) all the time, but body movement is not the type of mind over matter of interest here.

It seems that most people believe that they have some influence over physical objects or events. Gamblers attempt to influence the throw of dice in a game of craps or the spin of a slot machine. Nearly all golfers, including top professionals, "talk" to their ball, hoping it will listen and behave a certain way after they hit it. Belief that intention can somehow affect the world or bring good luck is pervasive. The question is whether this is possible or just a myth. The scientific method seems to provide the answer.

Experiments studying the effects of mental intention on the throw of dice have been going on for almost eighty years and provide conclusive evidence that such effects are real. A meta-analysis of 148 independent experiments indicated that the odds that mental intention does not influence dice are statistically one in a billion.[10] Interestingly, when no intention is applied to the throw of dice, researchers found that the hit rate was exactly that predicted by chance. Do these results prove that mind can affect the throw of dice? One would have to conclude, based on the high quality of the scientific controls employed and the reproducibility of experiments by more than fifty investigators that the answer is yes.

Today most of the studies investigating psychokinesis involve experiments trying to affect random-number generators (RNGs).[11] Such experiments are easier to control and can involve trillions of randomly generated bits (ones and zeros), which can be recorded automatically and with ease—in fact, the whole test can be automated eliminating operator bias or error. The effect also tests mind-matter interaction on a very small or quantum level, which appears to be how mind fundamentally affects matter. Radin and coworkers performed a meta-analysis of 152 published reports from 1959 to 1987 describing 597 experimental studies in which persons were instructed to try to make the RNG generate an excess of either zeros or ones; and another 235 control studies in which no mental intention was used. Overall, the experimental studies generated a 51% hit rate while the controls were exactly 50%, as expected. Odds that such results would occur by chance are one in a trillion.[12]

Of interest are the observations that being close to the RNG had no significant advantage over being at a distance; the intention of two or more people generally had a greater effect than a single individual; time was also irrelevant to the results for the same bias in the results were obtained whether a person's intention was applied before, during, or after the data was collected by the computer.

The skeptics—it's all just pseudoscience

Naturally, skeptics are quick to dismiss the veracity of the data indicating that ESP is real. The very existence of ESP would put a dagger through the heart of the physicalist's doctrine that mind is localized in the brain. There is probably no other phenomenon that if proven true by the scientific method would do more to upset the materialists' worldview. Hence it is not surprising that this group comes down hard on ESP research and has attempted to disparage the data supporting it. Take for example the following statement from Sean Carrol, physics professor at the California Institute of Technology and self-proclaimed atheist:

> The only problem is, parapsychology is not science. It's pseudoscience. From a completely blank-slate perspective, one can certainly pose scientific questions about whether the human

mind can tell the future or read minds or move objects around without touching them. The thing is, we know the answer: no. The possibilities have been investigated and found wanting; more straightforwardly, they would violate the known laws of physics. Alchemy was science once, but it's not any more. Not all hypotheses are equally worthy of our respect and attention; sometimes we learn that a particular idea doesn't work, and move on with our lives.[13]

We might wonder what laws of physics Carrol believes ESP violates, but he may really mean that the existence of ESP and other paranormal phenomena would violate his preconceived notion of how nature "ought" to work.

Marcello Truzzi, sociologist and magician, was one of the cofounders for the Committee for Skeptical Inquiry (CSI). The "mission" of CSI was to debunk anything that they considered pseudoscience—such things as acupuncture, astrology, chiropractic, ESP, healing, homeopathy, near-death experiences, reincarnation, etc. In other words, anything that bordered on the paranormal or was not currently accepted by "mainstream" science. Truzzi, unlike most of the members of the organization, realized that an open-minded approach to skepticism was better than a dogmatic approach. He coined the word "pseudo-skepticism," by which he meant a type of false skepticism where one pretends to have a questioning attitude, but in reality one's mind is hermetically sealed against whatever may threaten one's preconceived idea of what is true. Furthermore, he felt a true skeptic must be genuinely curious enough to look at the data from an agnostic point of view before rendering an opinion. At the same time, the skeptic needs to acknowledge that the burden of proof is not solely on those who make extraordinary claims but also on anyone who claims that a positive result is incorrect for some reason. The critic is also making a claim and thus must bear a burden of proof. Skeptics that do not recognize these truths can be labeled dogmatic or pseudo-skeptics—they see no reason even to consider the evidence since it cannot possibly be valid.

Only a year after he helped found CSI in 1976, Truzzi left the organization because he wanted the CSI journal to have a "balanced" approach to such things as ESP and print articles by researchers in favor of psi, but the majority of the members would have none of it.

In fact, scientists who do psi research are applying their time and talent to their work not because of any preconceived idea about reality but because of a real scientific curiosity, and they are not immune to criticism from those who think all such investigations are phony. They typically take extraordinary care to maintain the highest scientific controls and procedures—normally much stricter than other psychologists do.

The list of criticisms of ESP research by skeptics is large, but parapsychologists have answered these criticisms one by one.

- *The studies are not well controlled.* If better controls were used then only chance results would be obtained. Actually, the data is contrary to this. As studies were increasingly controlled and refined in face of criticisms, the results did not become less significant but actually increased in statistical significance vs. random chance.[14] In addition, some researchers, being sensitive to such criticisms, consulted with professional mentalists and magicians who could find no flaws in their methodology. Interestingly, it has been shown repeatedly that skeptics of ESP do worse in tests than people who believe or are at least open to the idea.
- *The studies are not reproducible.* Actually, the studies are reproducible. Many studies have been independently replicated. For example, presentiment studies were replicated by an independent investigator using his equipment.[15] Moreover, there have been hundreds of ganzfeld telepathy experiments, all getting positive results.
- *Nonconforming data is not reported.* This is the so-called "waste-basket" hypothesis that only studies showing positive results are published, which creates bias in the data. If the unpublished negative results were included in the data then the meta-analysis would not show positive results. The problem is the statistics do not support this criticism. For example, there would have to be fifteen unreported trials showing negative results for each reported ganzfeld study. For the ESP card experiments it would take about thirty thousand unsuccessful studies out

of thirty-four reported studies to reduce the odds to chance.[16] Hence, considering the time and effort it takes to conduct even one study, selective reporting cannot possibly explain the positive results obtained for telepathy, clairvoyance, precognition, and psychokinesis.
- *Results are fraudulent or faked.* Investigators really want to believe that psi is real and either consciously or unconsciously manipulate the data in order to achieve their expectations. While there have been a few cases of fraud or error, this is no reason to dismiss the vast majority of data. And while it is true that skeptics often get poorer or negative results when they attempt to replicate experiments of scientists who study psi phenomena, accusations that seasoned scientists, most of whom work at well-respected centers of higher education, would purposefully misreport their findings is patently ridiculous. Such scientists are fully aware of the controversial nature of their research and have every reason to want to protect their professional reputation. Therefore, they take a rigorous approach to insure that their work is of the highest standard possible—much more so than typical psychological studies. Furthermore, even if one or two studies out of two hundred were biased or flawed, it would not explain the other 198 studies, which gave positive results. In addition, statistical outliers are typically not included in a data set undergoing a meta-analysis.
- *Science is constantly explaining the supernatural in natural terms.* Throughout history, science has discovered that many phenomena that were earlier thought to be paranormal are actually explicable in natural terms. Of course, this is a phony argument since it says nothing about whether ESP is real, only that other phenomenon, originally thought to be mysterious, were later explained by science.
- *Psi effects are tiny, irreproducible, and can only be shown to occur using statistics.* This is an invalid criticism since the same can be said of most studies in the "soft" sciences. In addition, results are reproducible and in some studies, e.g. using intention imprinted devices (see below), the effect is so robust that

statistics are not required to demonstrate the scientific validity of the effect.
- *There is no theory to explain psi.* This is simply untrue. Psi phenomena can be simply explained by the existence of Cosmic Mind, which can also be called the One Mind or collective unconscious or "Mental Internet." These are names used to describe a collective mind, which is nonlocal in nature and represents the wholeness that is at the foundation of reality.

Evidence from intention imprinted electrical devices (IIEDs)

Suppose a scientist wants to know if an electronic device can be influenced by a person's intention. First, he buys two identical pH meters. One of the machines is placed next to a highly qualified meditator who attempts to imprint the device with the intention that it gives low pH readings. The devices are then wrapped in aluminum foil and shipped approximately 2,000 miles to a laboratory where the actual target experiment is conducted by others. There a sample of purified water is prepared with a small amount of added potassium chloride (to increase conductivity of the water). Half the sample is put into a clean beaker, the pH electrode placed in the solution, and the meter turned on. The other half is measured similarly with the other meter. The unimprinted meter reads a pH of 6.0 and the imprinted meter reads 5.0. The researcher repeats the experiment, but this time the meditator is asked to imprint one of the meters with a high pH. Interestingly, the imprinted meter reads one pH unit higher than the control meter.

Experiments like this have actually been done. The results are both robust and reproducible. Beside pH, such things as the UV spectrum of DNA-water solutions, the thermodynamic activity of enzymes, the growth rate of fruit-fly larva, the temperature and conductivity of water, and the breakdown voltage of a gas-discharge tube[17] have been found to be altered by an imprinted device.[18] All these effects are reproducible in the direction intended and significant to the point that no statistical analysis of the data is required to demonstrate that the effects are real.

Stanford physics professor William A. Tiller along with his collaborators Walter Dibble and Michael Kohane have performed hundreds of similar experiments and have concluded that there can be no doubt that devices imprinted by an experienced meditator affect physical reality. Their studies using strict laboratory controls and protocols are summarized in their book, *Conscious Acts of Creation: The Emergence of a New Physics*.

Other psi studies

The Princeton Engineering Anomalies Research Laboratory (PEAR) has been studying the interaction of human consciousness on sensitive physical devices, systems, and processes for over thirty years. For example, PEAR published the results of a seven-year study in which a group of volunteers attempted to influence random-number generators across millions of trials. The observed effects were small (about one tenth of one percent), but over the databases, they compounded to statistically significant deviations from chance (one in four thousand).[19]

In another PEAR study human operators attempted to bias the output of a variety of mechanical, electronic, optical, acoustical, and fluid devices to conform to pre-stated intentions—all without recourse to any known physical influences. Under control conditions, all of these sophisticated machines produced strictly random data, but when human operators applied intention to the devices, the experimental results displayed changes that could only be attributed to the intention of their human operators.[20]

PEAR conducted over 650 remote-viewing trials. The protocol required one participant, the "agent," to be stationed at a randomly selected location at a given time and to observe and record impressions of the details and ambiance of the scene. A second participant, the "percipient," located far from the scene and with no prior information about it, tried to sense its configuration and character and to report these in a similar format to the agent's description. Using a numerical-scoring method to evaluate the accuracy of percipients' subjective descriptions and the physical targets, they concluded a probability against chance of three parts in ten billion. The data confirmed the hypothesis that there exists an unexplained mode of information acquisition that is independent of both the time and distance between the percipient and the target.

A program conducted by the Stanford Research Institute (SRI) during the 1970s was somewhat similar. The study was funded by the US Department of Defense to determine whether remote viewing could be used for espionage purposes. In these experiments, a talented or gifted viewer was asked to sketch or describe a target facility. In some studies, the sender visited the site and attempted to send mental images of the site to the viewer telepathically. For testing purposes, a secret facility in the United States was sometimes used, and for espionage purposes, the viewer was asked to describe a hidden facility in a foreign country in order to enhance and corroborate other intelligence about it. In 1976, SRI researchers Puthoff and Targ published their positive results using primarily one highly gifted viewer (Ingo Swann).[21] Their work has been criticized by skeptics for lacking sufficient controls. However, in 1988 May and colleagues analyzed all the psi experiments conducted at SRI over a sixteen-year period, most of which were remote viewing tests. The data set consisted of over 150 experiments and twenty-six thousand trials. Their statistical analysis indicated odds against chance of 10^{20} to one.[22] The program ended in 1994 after twenty-four years at SRI and the Science Applications International Corp. (SAIC), a defense contractor, and twenty million taxpayer dollars. It is not known whether the CIA or other US intelligence agencies currently utilize remote viewing as part of their intelligence-gathering procedures. Nor is it known whether Russia or any other country currently utilizes psi to gather intelligence.

A RNG study dubbed the Global Consciousness Project is currently being directed by Roger Nelson at Princeton. It began in 1998 and utilizes a worldwide network of RNGs that collect randomly generated bits along with their time stamp and send the data to a server at Princeton. Data from this ongoing study show that there is a significant spike in nonrandomness coinciding with major global events, such as the death of Princess Diana, Y2K, 9/11, and the funeral of Pope John Paul II.[23]

Many other globally important events caused significant spikes in the network of RNGs, suggesting that the collective mind of humans on earth have a small but reproducible effect on what is known to be a random process.

Several experiments have shown correlations between the electroencephalograph (EEG) traces of two volunteers. In such experiments, one person (the sender) is subject to a regular stimulus such as a flashing light, and the EEG of the distant receiver is observed to see if it jumps at the same rate as the brain of the sender. Results indicate that there is a significant and reproducible correlation for paired individuals and no correlation when there is no pairing.[24]

Carefully controlled scientific studies of mediums conducted by Dr. Gary Schwartz at the University of Arizona strongly suggest that they can obtain accurate information about people, places, and events.[25] The mediums normally attribute this to communication with discarnate entities, but they could also be obtaining the information by ESP. Schwartz also placed a charge-coupled device (CCD) semiconductor in a light-tight box in a dark room and recorded luminous discharges that he attributed to luminous beings. However, this could also be explained as a man-machine anomaly in response to human consciousness.

History is replete with stories of people sporting superpowers or yogis demonstrating siddhis such as levitation. Some of these individuals, for example Jesus, are thought to have been born with these abilities while others (e.g. Buddha) developed their extraordinary abilities through intense spiritual practices—such as yogic exercises and meditation. Dean Radin provides persuasive experimental evidence for the existence of supernormal or psychic powers in a few special individuals. His studies are presented in his book *Supernormal: Science, Yoga, and the Evidence for Extraordinary Psychic Abilities.*[26]

Do animals have a "sixth sense?"

Throughout history, there have been reports of animals detecting hurricanes, earthquakes, tsunamis, and volcanic eruptions before the event happens. Examples include dogs, cats, and other animals acting strangely—birds taking flight and rats leaving buildings before the ground begins to shake. This mysterious ability may allow some animals to sense geophysical changes in the earth before they happen.

For example, it was reported that before the giant tsunami hit the coasts of Sri Lanka and India on December 26, 2004, elephants screamed and ran for higher ground, dogs refused to go outdoors, zoo animals rushed into their shelters and could not be enticed to come out, and flamingos flew to higher ground.[27] At the hard-hit Yala National Park in Sri Lanka, stunned wildlife officials reported that hundreds of elephants, leopards, tigers, wild boar, deer, water buffalo, monkeys, and smaller mammals and reptiles had escaped to higher ground unscathed.[28] Surprisingly few dead animals were found at the park.

There has been at least one example where authorities successfully forecast a major earthquake based primarily on the observation of the strange behavior of animals. In February of 1975, a 7.3 magnitude earthquake hit Haicheng, China. Haicheng had approximately one million residents at the time of the earthquake. Chinese officials ordered that the city of Haicheng be evacuated about twelve hours before the earthquake struck, believing that there was a high probability of an earthquake occurring. The prediction was based on widespread reports of unusual animal behavior. Only a small portion of the population was killed or injured because of the evacuation. It was estimated that the number of fatalities and injuries could have exceeded 150,000 if no evacuation warning had been given.[29]

Many dog owners have concluded that their dog knows they are coming home, and some animals seem to be able to sense that a person is about to die or have a medical emergency. An example of this was reported in the July, 2007 issue of the *New England Journal of Medicine* about a cat named Oscar that seemed to be able to predict the deaths of patients in a nursing home in Providence, Rhode Island. Just before patients died, Oscar would sit down by their beds and would become very upset if forced out of the room before the patient died. The article cited at least twenty-five successful predictions of impending death by the cat.

Dogs have been trained to give warning of an impending epileptic seizure in people prone to this condition. They are specifically trained to give warning to their owners so that they can take appropriate precautions before the seizure strikes. Whether the dog smells something from the person that provides a clue to an impending seizure or just uses its "sixth sense" is unknown. Dogs are also known to be able to detect cancers in people, even at an early stage. It has been postulated that it is their keen sense of smell that allows them to detect malignancies in humans, but this is only speculation. Another explanation for the uncanny ability of animals to sense things beyond the purview of the five standard sense organs is that they tap into nonlocal mind.

Many accounts of pets finding their way home from great distances and to unfamiliar locations have appeared in the scientific literature and popular media. One such example is the story of Bobbie, a female collie who was traveling with her family from Ohio to their new home in Oregon. During a rest stop in Indiana, Bobbie ran off and could not be found. After many hours of searching, the family gave up the search for Bobbie and continued to Oregon. After about three months, Bobbie turned up at the doorstep of their new home in Oregon. Bobbie had never been to Oregon.[30]

The homing ability of some animals is a confounding mystery to science. One can postulate a sensory explanation for such behavior, including the utilization of several senses simultaneously, but the simple explanation for such behavior and ability is that the animals are obtaining nonlocal information in a manner similar to human ESP.

Summary

Historically, religious beliefs often clashed with science. Scientific discoveries were sometimes dismissed by powerful religious institutions when the scientific findings contradicted Church doctrine. For example, Galileo was forced by the Catholic Church to renounce his findings that Earth revolved around the Sun or face imprisonment. Because of the tension between science and religion, science had to disconnect itself from anything religious or spiritual in order to proceed with its task of describing reality in a nonsuperstitious, nonmetaphysical, and fact-based manner. The spirit behind this movement in science is mostly positive, but it has caused science to swing to the other extreme—physicalism (aka materialism, reductionism, etc.), which essentially denies that there can be an alternate, nonmaterial-based explanation for phenomena that do not fit the reductionist doctrine. As a result, attempts to study or explain phenomena in nonphysical terms have been openly suppressed, ruthlessly criticized, or simply dismissed as pseudoscience by many of today's scientists.

Throughout the history of science, we can find examples of scientists who doggedly held on to preconceived attitudes about natural phenomena even when there was clear evidence contradicting their belief. For example, in the latter half of the eighteenth century the great majority of scientists were convinced that stones (meteorites) could not fall from the sky because from theoretical considerations they felt it was impossible for stones to form in the Earth's atmosphere. It was inconceivable to scientists at the time that there existed any solid matter in the heavens apart from comets, planets, and stars. Most reports of falling stones were simply rejected out of course and even those scientists that gave credence to such reports adhered to the hypothesis that they did not originate from space but instead were such things as volcanic ejecta or stones hit by lightning. This was despite the fact that there existed extensive collections of meteorites in several museums at the time.

Specialization in science has made it harder for people to consider an alternative to materialism. Science today is incredibly specialized. In order to be at the forefront of scientific research one must normally focus their energy on a very narrow slice of the pie. As a result, most scientists today have little inclination to study what might be considered philosophical questions about the nature of reality (metaphysics). It is natural to go with the majority opinion that matter is the ground substance from which reality springs. To suggest otherwise leaves one open to scorn and possible harm to their professional reputation.

The public is more open to the idea that psychic phenomena are real but still tend to follow the lead of scientists. In one survey, the percentage of people that believed that psychic phenomena were real was 68 percent. This is considerably higher than the percentage of scientists who believed—only 6 percent in the same survey of members of the

National Academy of Sciences.[31] We might suspect that scientists are simply better informed than the public about the paranormal. However, there is no evidence to support this. More likely, they are simply more biased and less open to new ideas, since they are devoted to a belief system that denies the possibility of ESP. It is natural that many people who are not scientists follow the lead of well-known scientists who appear in the popular media, most of whom embrace the doctrine of materialism.

In addition, the acceptance of psychic phenomena may require a more expanded consciousness. It is likely that accepting the reality of psi requires that a person have a wider, more universal or spiritual outlook on life. An analogy would be a blind person who one day miraculously gained sight. That person might begin to perceive a far richer, more beautiful reality with its promise of a new meaning and purpose of life, opening up a new realm of almost infinite possibilities. Similarly, someone who is mentally fixated on the material realm may find it difficult to appreciate or tune into a nonphysical ontology. As a result, no amount of evidence is going to change their worldview—a change in how reality is perceived or experienced may be a prerequisite.

16

The Nonlocality of Quantum Mechanics

The elementary particles are certainly not eternal and indestructible units of matter; they can actually be transformed into each other. The world thus appears as a complicated tissue of events, in which connections of different kinds alternate or overlap or combine and thereby determine the texture of the whole.

—Werner Heisenberg

Entanglement

A SCIENTIST WORKING AT a university near San Francisco wanted to test her theory that small objects in the universe could be connected in a mysterious way that was not limited by time or space. To test this hypothesis she designed two identical lottery machines that each held forty ping-pong balls. Half the balls were painted white and the other half were painted black. The balls were mixed randomly by

THE NONLOCALITY OF QUANTUM MECHANICS 141

a fan and two counter-rotating paddles. When a door to a tube was opened, twenty balls were blown into the tube, one by one. Control tests indicated that both devices produced random sets of black and white balls. After transporting one of the machines to a colleague in New Orleans, both machines were started simultaneously and the two scientists recorded the order of black and white balls produced by the machines. After comparing the lists of 20 black and white balls, it was found that they were in identical order, i.e. the first ball in both machines was the same color, and all subsequent balls matched as well. When the scientist calculated the odds of this happening by chance, it was less than one in a million runs.

This result indicated that there had to be some sort of connection between the black and white balls over the distance of nearly two thousand miles separating them. Although subsequent investigation could not find any known way by which the balls could be "talking" with one another, it would take only a fraction of a second for signals moving at the speed of light to travel between the two machines.

In order to eliminate light-speed communication, the scientist decided to repeat the experiment by placing the distant machine on Mars where it would take thirteen minutes for signals moving at the speed of light to travel between the machines. However, the results were the same as before—the two sets of data were again perfectly correlated. This second experiment demonstrated conclusively that local signals traveling at the speed of light could not be responsible for the result. According to the laws of physics, there was simply no explanation for how the two randomly generated black-white data sets could be correlated, yet they clearly were.

Of course, this experiment has not been done, but it does illustrate the results obtained by scientists who have investigated the connection observed between small particles such as photons, electrons, atoms, and molecules. Physicists have a name for this connection: entanglement. It was predicted by quantum theory and many scientists, including Albert Einstein, believed it was impossible and concluded that quantum mechanics had to be flawed. Physicists still use terms like "weird," "crazy," "bizarre," and "inexplicable" to describe entanglement. It demonstrates a level of reality that is governed by nonlocality—that is, a connection that is not dependent on either time or space. For entangled particles, a change in one is instantly communicated to the other, and the connection is not dependent on the distance between them.

Entanglement occurs when two or more quantum particles are "born" together in the same process. It was first proven to exist in the 1980s, but since then it has been confirmed by hundreds of experiments. Today quantum entanglement is being used to create what may be the next breakthrough in computer technology—quantum computing. Quantum computers use quantum bits (qubits) instead of bits. The quantum information encoded by a qubit contains information about the quantum state of the qubit—not just whether it is one or zero; but because a qubit can be a superposition of many states, the power of such computers can theoretically be orders of magnitude greater than that of classical computers in use today. Although the development of such "supercomputers" is still in its infancy, researchers at IBM have successfully built a prototype processor having fifty qubits. While fifty bits for a normal computer is equivalent to seven bytes and could not even code for the word "computer," in quantum physics one requires 2^n bits to describe the system completely. For just fifty qubits, this could theoretically provide 10^{12} bits (one hundred terabytes) of data, which is equivalent to ten times the print collection of the U.S. Library of Congress.

A quantum computer with its entangled qubits, which can be either one, zero, and states in between, is able to make many calculations at the same time, expanding its potential computational power millions of times over that of today's most powerful supercomputers, which can only make a single, albeit very rapid series of computations. For example, think of a single rat placed into a complicated maze with hundreds of dead ends and only one way out. It might take the rat many hours to find its way out as it tries numerous paths, only to be blocked most of the time. Now consider putting a hundred rats into the maze at the same time. Surely, one of the rats, by chance will find the elusive escape route in a short time. In the future, quantum computers will surely be used to greatly enhance robotics, artificial intelligence, and solve many complex problems such as weather and financial forecasting that are not possible using today's classical computers.

Quantum communication is another technology made possible by entanglement. Recently, Chinese researchers were able to use a space

THE NONLOCALITY OF QUANTUM MECHANICS   143

laser on a satellite to send entangled pairs of photons to two sites in Tibet some 1200 km apart.[1] This experiment was an important first step in demonstrating quantum communication, which uses entangled photons to encode information in such a way that it would be impossible to break the encryption. Normally if you want to send a coded message between two people, you must give them both a secure key that allows them to translate the message. At the same time, you must protect that key from any nosy third parties who are trying to spy on the conversation. A complex quantum key, shared via entangled particles, would do the trick, because if a spy tried to steal the code-breaking information, this would disrupt the entanglement making it useless for the intruder. In addition, it would inform the intended recipient that there was an attempt to intercept the message. Hence the beauty of quantum communication is that the integrity of the data sent is protected by the laws of physics and thus there can be no higher level of security.

Quantum teleportation is similar to quantum communication in that quantum information, such as the exact state of a photon, electron, ion, or atom, is transmitted from one location to another. It is another well-established example of quantum entanglement. It differs in that quantum teleportation provides a mechanism for moving qubits from one location to another without physically moving the underlying qubit particles. Quantum teleportation can take place when there is previously established quantum entanglement of particles at the sending and receiving locations, and the information about the particles is sent by way of a "quantum channel" to the receiving station from the sending station. In the process of transfer, the information carried by the particle at the sending station is destroyed. While the word "teleportation" conjures up images from *Star Trek*, it cannot be used to transport material objects—only information, and it would take enormous technological advances before this quantum information could be used to assemble even a simple object.[2]

The connection that exists between distant but entangled particles without any constraint of time and space is perhaps the greatest mystery posed by quantum mechanics. How can two photons, one in our galaxy and another in a distant galaxy, be in instantaneous contact with one another in such a manner that if one is observed to have a horizontal polarization the other instantly knows this and assumes a vertical polarization? This sort of mysterious connection between quanta that are entwined from their birth is no longer speculative but is a fact of nature. The best and perhaps only logical explanation for the phenomenon of quantum nonlocality is that all quanta are connected, inseparable parts of a wholeness that underlies everyday reality.

Entanglement has been shown to occur not only for small particles but also for complex systems. There is even evidence that it occurs in cells of our bodies. In addition, there is no theoretical limit to the extent of entanglement. Experiments indicate that quantum entanglement grows exponentially with the number of particles involved in the original quantum state.[3] Since the universe is believed to have begun in a singular state that erupted in a massive expansion (Big Bang), it is believed by some scientists that on a basic level the universe consists of a vast web of particles that remain in contact with one another throughout all time and space.

What quantum mechanics implies about the nature of reality

The study of quantum physics leads inexorably to the conclusion that the quantum world exists in both temporal and spatial nonlocality. By temporal nonlocality, we mean that the past, present, and future coexist in the now, and that time does not flow linearly from the past to the present to the future. For example, in some experiments it is observed that an effect precedes the cause. Spatial nonlocality means that quantum particles cannot be placed with certainty at any given set of coordinates. They are described mathematically by a wave function, which provides only probabilities of where they will be found when observed or measured; and the wave-like quanta are not localized in space, which means there is a finite probability that they can be found anywhere in the universe.

One of the classic experiments of modern quantum physics that reveals the nonlocality of space and time is the dual-slit experiment. In this experiment, light (photons) or electrons are passed through two slits that are very close together. Both electrons and photons have wave and particle characteristics depending on how they are observed. Typically, the wave-particles passing through the slits interact as waves to form an interference pattern (alternating dark and light lines) as the waves either cancel or reinforce one another. This is exactly what is expected if light behaves like a wave.

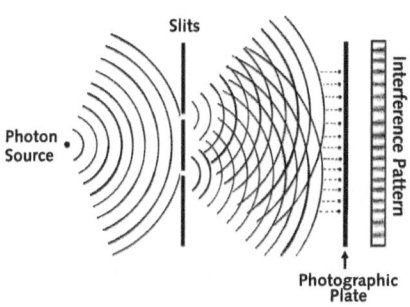

Dual-slit experiment. Waves from the photon (light) source pass through two closely spaced slits causing the wave fronts to interfere with each other producing alternating dark and light lines

However, light can also act like a flow of particles. If the intensity of the light beam is lowered sufficiently then only one photon at a time will pass through a slit of the apparatus. In this case, the interference pattern might be expected to disappear since it is created by the interaction of wave fronts from the two slits, and the individual photons arriving at the slits at different times should not be able to interact. However, the pattern persists even though the individual photons cannot split, each half going through a different slit and then recombining before they hit the detector. The interference pattern may be observed by exposing a photographic plate for many minutes or summing the response from an electron detector. The existence of the interference pattern under these conditions suggests that the so-called "well-defined" particles are behaving like a wave and individual particles are interacting with particles that passed through the slit at different times. That is, they exhibit both spatial and temporal nonlocality.

The dual-slit experiment can be modified in such a way that one of the slits is closed electronically after the photon has passed through a slit but before it strikes the detector. In this so-called "delayed choice" experiment, the interference pattern disappears and only a simple diffraction pattern is observed. Theoretically, if the photon is forced to pass through only one slit, it is forbidden from acting like a wave and reverts to its particle nature. The same is true when the slit is closed after the photon passes through one of the slits, except that this effect precedes the cause. This is a classic example of temporal nonlocality. The question is how does the photon that goes through the open slit know that the other slit is closed or about to be closed after it passes through and that it must go to a different location on the photographic plate? Somehow, the photon is aware that the other slit is not fully available to it and it acts like a particle instead of a wave.

Experiments have consistently demonstrated that the behavior of quanta is not simply determined by the conditions of the test. Their wave-like or particle-like behavior depends on the totality of the experimental apparatus and the intention of the experimenter. This outlines an important aspect of quantum mechanics—the observer and the observed system cannot be separated. The observer or his instruments are part of the system and influence the outcome of the observation. In other words, the act of observing alters or influences the system and this alteration occurs even when it is done after an effect. To summarize what we know so far about quantum nonlocality:

- *The wavelike or particle-like behavior of a quantum particle is determined by the decision of the experimenter.*
- *An effect can precede a cause or even occur before an apparatus is turned on.*
- *The two complementary aspects of quantum particles, i.e. wave and particle, cannot be observed simultaneously.*
- *An "act of measurement or observation" is required before a quantum system can manifest in physical reality or become "real."*
- *Quanta display spatial nonlocality. Quanta cannot be placed with certainty at any given set of coordinates. The equations of quantum mechanics only provide the probability of where quanta will be found when observed.*
- *Entanglement of quanta occurs under the right conditions and is nonlocal with respect to time and space.*

The quantum wave function— why it may reveal an underlying reality

In the 1920s, the pioneering work of Nobel Prize winning physicist Erwin Schrödinger provided a method for calculating the possible wave functions for a system. Therefore, the quantum wave function is also known as the Schrödinger wave equation. A wave function describes mathematically the properties of a wave such as water waves or vibrating violin strings. However, for quantum systems the wave function is not a wave in physical space, but a wave in an abstract "mathematical space." Although the mathematics are quite complex, in simple terms, the wave function details all the possible states that a particle or system may have and also gives the probability that it will assume any single state when "observed."

The Schrödinger wave equation provides an explanation for wave–particle duality. Before a quantum particle is observed and manifests in the "real" world it is best described as a wave. When it is observed or measured, it behaves like a particle. Observation apparently "kicks" the particle or system out of the realm of infinite potentialities into a specific state. When this occurs, it is called the "collapse of the wave function."

 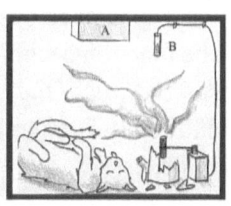

Since we cannot know the state of a particle before it is observed, quantum theory concludes it must be a superposition of all possible states. This condition is illustrated by a well-known thought experiment, which has been dubbed "Schrödinger's Cat." In this scenario the behavior of a quantum particle is linked to the life or death of a macroscopic entity—namely a cat. The cat is put in a box along with a trace amount of radioactive substance in $\underline{A}$ that might omit a beta particle every hour or so, along with a Geiger counter at $\underline{B}$ that is linked to a hammer that will break a vial of poison. There is no way of knowing when the substance will decay and hence whether the cat is alive or dead without looking in the box. Quantum indeterminacy would require that there exist a superposition of states in which the cat is half-dead and half-alive until an observation is made. This thought experiment illustrates one of the oddities of quantum mechanics—namely that a system can exist in two or more states simultaneously until the superposition collapses upon observation.

Since the wave function details possibilities, it speaks of a reality that is subtler than what we perceive as physical reality. This domain is not one of separate parts but one in which all possibilities coexist in a state of wholeness—everything being interconnected and interdependent. By necessity, this web of connectivity must permeate the entire universe. Hence, the quantum wave function describes a subtle realm of both temporal and spatial nonlocality. Below is a summary of the properties of this hidden realm.

- *It expresses wholeness.* The function represents a fundamental aspect of reality that exists at a deeper level than ordinary reality.
- *It is timeless.* The past, present, and future are meaningless when discussing this realm. It is only after the wave function collapses by observation or conscious awareness that an arrow of time comes into existence.
- *It is nonlocal.* It penetrates and surrounds ordinary reality and is not localized in any part of space but is all encompassing, everywhere at the same time. It is only after an observation that a "part" of reality represented by the function becomes localized in space.
- *It is a mathematical representation of the possibilities.* It determines the probability that any particular quantum possibility will become "real" (when observed).
- *Theoretically, all matter and energy have associated wave functions.* This includes the brain, the body, and the universe as a whole. The wave function represents the gestalts for these individual entities.
- *The domain of the wave function does not contain energy as such.* Instead, it is the underlying source of all energy. In a sense, it contains the potential for expression of almost infinite energy.
- *The collapse of the wave function does not require energy—just observation (consciousness).* This appears to be the mechanism

by which mind, which is nonphysical, affects matter (brain); conservation of energy is not an issue.

Interestingly the properties of the hidden domain of the wave function are identical to the properties of Cosmic Mind. The wave function serves as the mathematical underpinning of quantum mechanics and because its description of reality has been repeatedly verified by countless experiments, we must conclude that quantum theory is entirely consistent with the spiritual worldview but antithetical to the spatial and temporal locality of the materialist worldview.

Summary

Quantum mechanics and relativity theory are the cornerstones of modern physics. Einstein's relativity has been instrumental in enhancing our understanding of the cosmos on a large scale, and quantum mechanics has been tremendously successful in describing the interactions of matter and energy on a small scale. Although physicists have not yet been able to integrate these two great theories, they both point to a reality that is more subtle and unifying than the "everyday" variety we perceive with our sense organs.

The theory of relativity has provided astrophysicists with the mathematical tools needed to understand the birth and evolution of the observable universe. It is now believed that our universe began 13.8 billion years ago from a dimensionless point. Before the dawn of creation (the so-called Big Bang), there was no space or time. Somehow, the entire mass/energy of the cosmos emerged from this point creating four-dimensional space-time. According to the model, the point of origin of the universe is not at any specific location, but is essentially everywhere in the universe. In the first fraction of a second following the birth of the universe, it inflated much faster than the speed of light and then things collapsed into the slowly expanding universe we see today. The theory has been exceptionally successful in explaining many of the fundamental properties of the universe and the connections between mass, energy, gravity, and time. However, it indicates that the true nature of reality is four-dimensional but under normal circumstances we only perceive a three-dimensional slice of this reality. For us time flows from the past to the present to the future much like water flows under a bridge. However, the flow of time is illusory—created by the constant movement of our three-dimensional world through an unchanging four-dimensional space. We can neither conceive nor perceive the "big picture," which would be the timeless four-dimensional reality that Einstein called space-time.

Quantum mechanics indicates that physical reality is nonlocal. It says that quanta can be connected and that the connection is not affected by the constraints of time and space. It implies that beneath physical reality lays a domain that is best described as whole and timeless. Experiments have repeatedly indicated that quanta become "real" only after some form of observation causes the collapse of their wave function. In other words, nothing can emerge from this subtle realm of potentiality into what we call the realm of physical reality without an act of observation (consciousness). This implies that consciousness is "king," not matter. Interestingly, this requirement of quantum mechanics led to a famous exchange between Albert Einstein and Neils Bohr.

> Einstein: "Is the Moon there when no one looks?"
> Bohr: "Einstein, stop telling God what to do!"

Of course, Einstein's objection to quantum mechanics is moot. If matter is actually nonlocal as quantum mechanics indicates, it follows that mind, which is more subtle than matter, must be nonlocal as well. This conclusion is consistent with the enormous amount of evidence indicating that mind cannot be equated with brain. Behind mind is the observer/consciousness, which is nonlocal and is a reflection of the indivisible Cosmic Consciousness. Since the entire cosmos is a product of Cosmic Consciousness, the existence of the universe and its parts is independent of the consciousness of any living creature.

PART IV

THE PATH

The idea that we reincarnate only makes sense if there is a purpose or meaning to life on earth, and that meaning is to realize that we are a small part of a great and unfathomable Truth. The experience of the mystical wholeness or oneness of creation has been called various names, such as cosmic consciousness, enlightenment, liberation, self-realization, salvation, samadhi, moksa, and nirvana. Such realization might be momentary, but one becomes fully aware of life's purpose even as one goes about living a normal life. On the other hand, the ultimate union with God occurs upon death and there is no further need for rebirth since the soul becomes one with God. One analogy used to describe this unification is a drop of water falling into the ocean—becoming one with the ocean.

Just as there might be numerous pathways to the summit of a mountain, similarly there are many paths to attain self-realization. There can no more be a single way to the goal of union than there can be a single river that empties into the ocean—for each individual is different and a path that is right for one may not be the best for another. All we can say is that all the paths have some common features that I will try to outline in this part of the book.

17

The History of Unity

If those who lead you say to you, 'See, the Kingdom is in the sky,' then the birds of the sky will precede you. If they say to you, 'It is in the sea,' then the fish will precede you. Rather, the kingdom is inside of you, and it is outside of you. When you come to know yourselves, then you will become known, and you will realize that it is you who are the sons of the living Father. But if you will not know yourselves, you dwell in poverty and it is you who are that poverty.

—Jesus in the Gospel of Thomas

THE IDEA THAT THE entire cosmos is the manifestation of the Godhead, which means we are ultimately one with God, is a veracity that all religions point to underneath their seeming differences. This truth has been called "the perennial philosophy"—a term that was popularized by Aldous Huxley in a book by the same name. This philosophy of oneness may be expressed most succinctly by a saying in the Upanishads—*tat tvam asi* (thou art That); the immanent eternal self (atman) is one with Brahma, the Absolute Principle of all existence; and the goal of every human being is to discover this fact for themselves by discovering their Divine Self.

As mentioned in chapter one, Shiva was probably the first sage to teach a philosophy and path for obtaining unity with God. His was a tantric system that was based on moral behavior and spiritual practices—most importantly meditation.

Lord Krishna was another divine personality who lived in India about 1500 years BCE. He is best known from the Bhagavad Gita (Divine Song) which was a part of the Mahabharata—a major Sanskrit epic of ancient India. In the Bhagavad Gita, Krishna reveals himself as God to his disciple Arjuna. Krishna taught that this world is a creation of maya, the powerful force of illusion that creates confusion and distinctions. The way to overcome this powerful force of distraction, he says, is to surrender completely to the Lord. He taught his disciples meditation and other spiritual practices in order that they could transcend their ego and attain God-realization. Like Shiva, after his death Krishna became revered as one of the personified forms of God.

The Tao is "the path" or "way of life," and its founder, Lao Tzu, taught that to know God one must know oneself. Ultimately, the cycle of life, death, and rebirth is ended by unification with God. Buddhism has a similar concept, except that Buddha did not explicitly mention Cosmic Consciousness or God. In Buddhism, the end of suffering and rebirth is attained when one merges in the void (*shunya*).

About 150 BCE, Hinduism was revitalized by a great yogi and philosopher by the name of Patanjali. He compiled the Yoga Sutras, which delineated ashtanga yoga or the eight-limbed path of yoga. These eight limbs were the moral guidelines of *yama* and *niyama*, asanas (postures), *pranayama* (breath control), *pratyahara* (withdrawal of mind), *dharana* (concentration), *dhyana* (meditation), and samadhi (suspension of mind and absorption in the One). The meaning of the word *yoga* in Sanskrit is unification, and Patanjali argued that the goal of samadhi is attained when all the psychic propensities of the mind are suspended. In this state of suspension, the unit consciousness unites with Cosmic Consciousness.

Unity in the West

Jewish mysticism goes back to at least the sixth century BCE. Jewish rabbi-sages, after returning from captivity in Babylon in 538 BCE, started rebuilding the Temple in Jerusalem and helped reinvigorate Judaism with a new sense of mysticism.[1] These mystics may have been inspired by Ezekiel's vision of God's chariot. Later this Old Testament story of the prophet encountering God on a golden chariot spawned *Merkavah* (throne-chariot) mysticism, which focused on union with the Divine Being. Followers of this type of mysticism sought to recreate similar experiences and ascend into God's realm by fasting, ritual purifications, and repeating the names of God.

Jewish mysticism gradually morphed into Kabbalah. Reincarnation is a central tenet in Kabbalah as well as in Hasidism. It is known as *gilgul* (cycle of the souls). Souls are seen to go through human incarnations until they achieve divine union. Which body they associate with depends on the particular tasks they need to accomplish before they can ascend to spiritual union with God. Today in the more popular forms of Judaism, the whole concept of an afterlife is downplayed—with some rabbis suggesting that we live on only in the memories of others.

The Greek word *gnosis* refers to knowledge based on personal experience or perception. In a religious context, gnosis is mystical or esoteric knowledge based on direct knowledge of the divine. Gnosticism probably had its origins in Jewish mysticism, but it grew in popularity among Christian communities in the first few centuries CE. Gnostics believed that knowledge of self was knowledge of God. Salvation was attained through the knowledge of the Divine Self that lies within.

The Gnostic Christians believed they possessed the secret knowledge (gnosis) and practices that were taught by Jesus to only a few select individuals. The secret knowledge included the idea of "spiritual resurrection" as opposed to the physical resurrection of the orthodox Christians. Physical resurrection was defined as taking on a new and much improved physical body upon salvation on Judgment Day. This idea was bolstered by the belief that Jesus physically arose following his death as evidenced by his empty grave.

For the Gnostics their secret practices included meditation techniques to attain spiritual resurrection, which they defined as spiritual union with God. If this were not attained in one lifetime, then the soul would be reincarnated until spiritual union was attained.

Some authors have argued that Jesus was a Jewish mystic who taught the concept of divine union.[2,3,4] They cite specific passages in the Bible to support their hypothesis. For example, Jesus said, "the Kingdom of God is within you."[5] In his book *The Third Jesus,* Chopra points out passages in the Bible that suggest that the "Kingdom of God" is a state of consciousness—God consciousness to be precise. Hence, to attain the Kingdom of God is to attain union with God.

A contemporary model for spiritual unity

Perhaps the most detailed and scientific explanation for the role of humankind in God's creation and how to best attain unity was given by Prabhat Ranjan Sarkar (1922–1990). Sarkar was born in Jamalpur, India, and from a very early age he exhibited many extraordinary abilities, such as practicing meditation without learning it from a teacher; initiating much older persons in meditation; displaying great knowledge of languages, spiritual concepts, medicine, and various other topics, all gained without the help of teachers or books.

As a young student, Sarkar was considered by his teachers to be exceptionally bright but often bored with school and prone to daydreaming. At age eighteen, he left Jamalpur for Calcutta to attend the University of Calcutta but had to quit his studies two years later in order to support his family following the death of his father. For the next twenty-five years, he worked as an accountant at the railway headquarters in his hometown of Jamalpur while teaching the spiritual practices of tantric yoga to people that came to him for instruction. In 1955, while still working in the railway office, Sarkar formed the organization Ananda Marga (Path of Bliss) with the twin purposes of spiritual progress and social change. He accepted the yogic name Anandamurti, which means "embodiment of bliss." He began training missionaries (acharyas and avadhutas) to spread his teachings of "self-realization and service to humanity" all over India and soon throughout the world. The Ananda Marga organization eventually grew to become a large and multifaceted organization with members in over 130 countries, having different branches dedicated to the physical, psychic, and spiritual advancement of humanity.

Anandamurti became known as a spiritual teacher, scientist, philosopher, neohumanist, social theorist, linguist, artist, and economist. He wrote over two hundred books on various subjects, such as history, spirituality, sociology, education, Tantra, yoga, medicine, ethics, psychology, humanities, linguistics, economics, ecology, farming, music, and literature. He gave several thousand discourses and composed over five thousand mystical songs known as *Prabhat Samgiita* (Songs of the New Dawn).

He taught that creation was cyclical in nature. Cosmic Consciousness is transformed into Cosmic Mind and the material world in the extroversive phase (*saincara*) of the cycle, while living organisms arise in the introversive phase (*pratisaincara*) in which matter is transformed into living organisms. Human beings have an advanced mentality that reflects the pattern of Cosmic Mind and they naturally have a thirst for the limitlessness of Cosmic Consciousness. Anandamurti taught that the one characteristic that separated human beings from animals, their *dharma*, is their innate ability and desire to become one with God.

Anandamurti taught a system of yoga, meditation, and other spiritual practices to achieve all-around physical, mental, and spiritual development with the ultimate goal of spiritual union and the end of rebirth.

18

The Importance of Moral Behavior

The first step of sadhana (meditation) is Yama and Niyama (moral principles). Yama and Niyama constitute the very base of life of a spiritual aspirant, and serve as a beacon in the life of a spiritual aspirant, to dispel all darkness from the mind.

—Shrii Shrii Anandamurti

UNION WITH GOD IS impossible when there is substantial karmic debt to be repaid because the heavy burden of unburnt karma will surely lead to rebirth. Therefore, it is important to burn through our existing karma and not add additional karmic burden in this lifetime while we work to attain self-realization. This means that we need to observe moral precepts since the mind is dragged down by the weight of immoral actions and progress on the path to union becomes impossible. Hence it is no surprise that all the world's religions put great emphasis on thinking and acting morally.

Theoretically, all actions whether good or bad create reactive momenta (*samskaras*). Thus it would seem like an impossible task to burn through one's entire karmic debt before dying since there is sure to be one or more remaining reactive momenta in the mind at the time of death.

Theoretically, a person might have to suffer rebirth to burn just a few remaining *samskaras*. Fortunately, there is a way to go about life without creating new reactive momenta and it is possible to surrender your karmic debt at the time of death. However, to understand the spiritual principles that make such things possible it is important to consider the part of our mind that is chiefly responsible for our separation from God—the ego.

The ego trap

A baby is born innocent without ego. Early on, it has no sense of individuality or discrimination between self and nonself. The baby is oblivious to problems and attachments, totally accepting and trusting of its world. It knows only love, and it lives totally in the timelessness of the present moment. This egoless state of consciousness is one of bliss and unconditional love—something we might seek to experience later in life through intense spiritual practices. However, for a baby this perfect state of contentment is intermittent and does not last. Moreover, a child is not born as a blank slate. Each child brings its unique karma—traits that give the child its character and individuality. The mother instantly recognizes her baby from all others and knows its peculiarities. Even very early in a child's development, remnants from their previous life begin to color their individuality and self-image. However, their consciousness of self is undeveloped along with their memory, which for a young child is noncontinuous or sporadic. For adults only a few memories are retained prior to the third year, to say nothing of a previous life, but reactive momenta differ from memories in that they originate from an unconscious layer of mind and directly affect the physical body and mind.

In order to cope with the physical world, the child gradually learns the difference between self and nonself and begins to speak of itself in the first person. The developing ego imparts a sense of duality—the difference between me and it or me and you. The ego separates what is physically real from what is unreal. It helps the child organize its thoughts and make sense of them and the external world. This development is essential for the healthy growth of the child, for without it the discrimination and judgment needed for survival would be lacking. Gradually, as needs and desires arise that are not immediately fulfilled, pain arises and the beginnings of fear, anger, desire, attachment, and doubt emerge.

As the child develops mentally and physically it develops empathy, learns social rules of behavior, and forms friendships. The child's character is a product of both their environment and their genes. However, sometimes the child will take a life path that is unexpected and very different from that of its parents. This may be due to the need to express karma it carries from previous lives, which need to be manifested in this life.

Ego development continues into and beyond adolescence as the strong emotions of sexuality, love, dislike, and anger take hold. Ego development does not end here. Most people today believe that happiness is a result of their accomplishments. To be happy one needs a meaningful and well-paying job, a spouse, children, a nice home, and personal conquests. The ego's quest for achievement has no real boundaries because underneath the "I do" mentality of ego lies the limitlessness of Spirit. The thirst for the infinite when directed toward the material plane of existence is at the root of the problem with egoism, for even if one could acquire all the wealth, power, and fame of the entire planet it would not be infinite.

As the ego amasses layer upon layer of possessions, wealth, status, and power in its search for unlimited happiness, its burden grows and possessions do not bring as much happiness as before. A lack of fulfillment and a sense of dissatisfaction with life may develop. This can be healthy if it stimulates an interest in seeking union with God.

However, for many people this transition never takes place. They fail to learn of the limitless riches gained by transcending the ego and finding peace and happiness within. Carl Jung, in his book *Modern Man in Search of a Soul*, described the problems such people encounter as they age.[1] They try to hold onto the pleasures they experienced as a youth as they enter the second half of life. Naturally, they are accustomed to acquiring such pleasures in the external world, but their sense organs inevitably become duller and the motor organs weaker. Such people seek greater and greater stimuli in order to compensate and may fall prey to a "midlife crisis" or an autumn of life marked by discontent, dissatisfaction, cynicism, and unhappiness. No matter how hard one tries, it is impossible to reverse the physical effects of aging. Jung argued that to be mentally healthy as a person ages, they need to give serious attention to their psycho-spiritual development. Being freed from many of the mundane obligations of youth in their middle years, they have the opportunity to reap the incredible treasures that an introspective approach to life can provide—for true happiness springs from within, not from without.

It is the nature of ego to create separation between self and nonself, since it functions in the realm of "me" and "mine." In addition, the ego provides the sense of willfulness ("I do" feeling) that exists behind all that a person does, says, thinks, understands, or hears. It is at the root of selfishness, judgment, rejection, and separateness. Because it wants to control, it also wants to dominate the nonself, which includes other humans, animals, and nature. The ego is threatened by the idea of losing control. Death is the greatest fear of the ego, since death presumably brings loss of control and loss of individuality. Since these occur in self-realization, the ego is deathly afraid of becoming one with God. It will seemingly do everything in its power to prevent what it perceives as a loss of control. The fear of losing one's individuality at the time of death is a major hurdle to overcome since it can prevent the complete surrender required for union. This is the reason that it is very difficult to avoid rebirth unless one takes the time to perform spiritual practices. Hence, ego can be a great enemy for spiritual progress. Trapped in the domain of me and mine, people dominated by ego fail to experience the true happiness that lies beyond the material realm.

Ego has been likened to an umbrella blocking Spirit and preventing the mind from experiencing God's grace, which is continually raining down on us. However, ego is an indispensable component of a healthy mind, for without the "I do" engine of ego, one would become vegetable-like, unable to make any movements or decisions. Therefore, the goal of a spiritual aspirant is not to destroy their ego but instead to feel that God is acting through them. This is the solution to the problem of creating new karma since any action performed without the feeling of doership or ego does not create new reactive momenta.

The quest to transcend ego

Human beings have a deep connection to the Higher Self or Spirit. They have an inner voice that speaks to them in subtle whispers, and this voice is only quieted by the constant activity of the conscious mind or the stupor of unconsciousness. When an individual finally comes to the realization that there is more to life than the incessant quest for pleasurable experiences and achievements, then they inevitably start down the spiritual path toward union. Such a person might be called a "seeker." A seeker has acquired wisdom and realized that knowledge of self is knowledge of God. Such a person rejects the idea that extroversive activities bring lasting happiness. They have come to realize that true happiness comes from within.

A seeker has begun the search for the meaning of life and the attainment of self-realization. The major concerns of the ego are put aside as the seeker begins to give themselves to selfless service and develop feelings of universal love for humanity. The seeker has developed an understanding that there lies a far deeper and richer reality beyond ego attainment and the experiences of the sense and motor organs. In a sense, a seeker has crossed the threshold from the stagnant waters of a pond into the flow of a stream that will eventually lead to the ocean. Seekers are characterized not only by an interest in attaining self-knowledge but also by their practice of some form of introspection such as meditation. In their search for truth, such individuals become tolerant, nonjudgmental, loving, selfless, service minded, more in tune with their body and nature, healthier, and happier than they ever thought possible.

Not every step on the seeker's path is blissful since they are still encumbered with negative karma from the past and plenty of ego baggage to overcome, but the rewards they feel along the way offer convincing proof that they are on the right path. According to Deepak Chopra, a seeker is giving by nature and wants nothing in return, not even gratitude, being motivated by selfless love and compassion. Intuition becomes a trustworthy guide, replacing strict rationality; they catch glimpses of an unseen world as the higher reality and intimations of God and immortality appear. These signs are accompanied by growing enjoyment of solitude, by self-reliance in place of social approval, by stirrings of Being and willingness to trust.[2]

As a seeker progresses on the spiritual path to union, they will eventually experience a flash of samadhi or enlightenment. Suddenly they have the mystical vision that everything is a manifestation of God. At that moment, the illusion of separateness breaks down. For that person unity is no longer something taken on faith or belief—it is experienced. They have surrendered their ego if only for an instant and experience the world in its pure form—as manifest Cosmic Consciousness. Slowly the seeker is transformed into a person we can call a "seer." Such a personality feels love for every living creature and every particle composing the universe. They are completely open, no longer play psychological games, and are incapable of feeling any emotion except unconditional love for God and his creation. They live totally in the now and have access to the unlimited knowledge that lies in the collective unconscious or Cosmic Mind.

The seer is not attached to the fruit of their actions; they feel that their actions are those of God. They create no new karma and know with certainty that when they leave their physical body they will become one with God. A seer is completely free, living in a state of grace and indescribable bliss. It is natural for people to be drawn to such individuals and look up to them for advice and spiritual guidance. Knowingly or unknowingly all human beings crave the limitlessness that accompanies this, the ultimate phase of life.

Surrendering the ego is a technique designed to reduce our karmic burden and reduce our sense of separation and to increase our love of God. The technique is taught by all the world's great religions. For example, Jesus taught unconditional, self-sacrificing love for God and for humanity. He preached service, humility, and forgiveness. He is quoted as saying that one should turn the other cheek if someone slaps you,[3] and to love your enemies and pray for those that pursue, slander, and falsely challenge you.[4] He said that one must become like a child to enter the Kingdom of God, completely loving and trusting.[5] It is obvious that Jesus knew that the greatest barrier separating us from God is our ego. His teachings, if followed with real awareness, would lead to a life dedicated to serving humankind, and the burning of our remaining reactive momenta. Such selfless service is a great tool for diminishing the ego's control over our life—allowing us to see through the illusion of separateness from God.

If an action is performed with the ideation that God is performing the action, then the mind suffers no reaction to that action. The thought can be, "I am the machine, and God is the machine operator." The Sanskrit term for this type of ideation is *madhuvidya*, which simply means ascribing godhood to every living organism and object. It is a practice that allows one to carry on a normal worldly life and not create additional karma. For example, when walking down the street one might try to see other people as manifestations of God. By serving them as though one were serving God, one's exterior and interior is filled with cosmic bliss and all afflictions are extinguished.

Summary

God is the ultimate Good, and if we could experience the egoless state, which is a characteristic of enlightenment, we would feel intimately connected to him and bask in the ecstasy of his being. In reality, humans are no more separate from the Source than rays of sunlight are separate from the sun, but ego gets in the way by creating the illusion of separateness. Ego obscures Spirit, covering it with layer upon layer of "I am so and so," and "I do such and such." Ego is the source of our arrogance and suffering, and to believe that it should be empowered is the epitome of ignorance.

Most people are unaware that there are practices that can speed up their spiritual growth and end the nightmare of constant rebirth. They continue living life on the surface of their being. Inevitably, they learn from the struggles of life experiences that it is necessary to subjugate the ego to attain happiness. Often it is hardships and feelings of unhappiness, dissatisfaction, pain, and suffering that cause people to change the direction of their lives. Hence suffering can be a blessing in disguise, and great suffering can beget great growth. This process entails the burning of negative reactive momenta, which brings mental anguish, pain, and suffering, but in the process the ego becomes subtler, yielding more of the sentient "I feeling." For some people these experiences are what are needed to turn them away from the false promise of ego attainment—"the material path"—to the path of unity.

Western scientific reductionism and religions do little to invalidate the myths of separateness and materialism. As a society, we pay a high price for subscribing to these dogmas. The majority of ordinary people live by the principle of separateness. Living only on the lower planes of existence, they identify themselves with their bodies and lower minds. In this state of ignorance, they see themselves as separate from the world and from other human beings. They erect social and psychic barriers between themselves and others—barriers such as nationality, race, gender, religion, and economic status.

19

The End of Rebirth

If I go into the place in myself that is love, and you go into the place in yourself that is love, we are together in love. Then you and I are truly in love, the state of being love. That's the entrance to Oneness. That's the space I entered when I met my guru.

—Ram Dass

MYSTICS DESCRIBE ENLIGHTENMENT AS awakening to a clear understanding of the unity that is. The word "awakening" is useful because it is analogous to awakening from a dream. While dreaming one does not question the reality of the images that are witnessed, even though they may be quite bizarre. Upon waking up one will immediately dismiss the dream as unreal. Similarly, the illusion of separateness or non-unity appears real to us, but upon awakening to an enlightened state, we know that this state of consciousness was illusory.

Furthermore, the mystics state that in the awakened state one loses the sense of a separate selfhood. It is identical to the state of consciousness of the seer. Everything is seen and felt as the One and there is an end to suffering. A natural state of bliss arises from the non-separateness that occurs when one lives totally in the here and now. One continues to experience life with its joys, pleasures, pain, and love, but these experiences are not resisted by the illusory "me." In other words, one's experiences are witnessed by the mind but have no effect on it.

Mystics describe our bondage by time, place, and person as that of a prisoner who is confined to a dark and drab jail cell. Worse yet, our confinement includes chains—both iron chains of bad karma and golden chains of good karma that doom us to be reborn in the cell. Like the prisoner, we know that outside the prison, there is the bright light of Spirit and the indescribable bliss of being one with God, but we have no idea how to escape from our predicament.

Fortunately, the mystics have found a path leading to liberation from bondage. They describe it as a gradual transformation that takes great effort. If we believe the mystics, then even if we have unfailing faith, adhere to moral principles, and accept God into our heart, our salvation is still not guaranteed. Lacking the spiritual training to see God in every person and everything in the universe, our mind at the time of death is apt to be dragged down by the weight of our unfulfilled desires, attachment to the world of matter, denial of our true being, and fear of losing our individuality in the infinity of Spirit. Unwilling to completely surrender our being to God, we naturally choose rebirth.

In this chapter, we will explore the practices that have been tested, refined, and proven to work for progress on the spiritual path. When performed with diligence these spiritual practices should gradually transform the mind from its infatuation with matter to the nonlocal experience of Spirit.

The power of prayer and selfless service

Prayer can be of value for moving forward on the path to union. The lower forms of prayer in which one asks God for some favor such as "God, please let me become rich" or "God, do my enemies harm" have no value. God already knows everything about us; he knows exactly what we need. Selfish requests will not be honored. A higher form of prayer is to thank God for what we have. This too is mostly a waste of time if the thanks are in the hope of receiving some favor from him in return. We may feel that we are being humble by thanking God, but he does not need any thanks from us. He loves all of his created beings unconditionally.

The highest form of prayer is devotional prayer in which a person asks God for emancipation. This is different from seeking a favor from God, because God's purpose in creating humankind is to make the unit beings free. This is the wish of God, and everything in this creation is directed toward that end. Thus the only valuable form of prayer is devotional prayer. Since the mind takes on the qualities of the object of its concentration, if one constantly calls on God, their mind will become filled with love for him.

Even a simple prayer can be used. For example, in the nineteenth-century religious classic *The Way of the Pilgrim*, the narrator, a wandering hermit, attains a high state of spiritual awareness and bliss by ceaselessly repeating the Jesus Prayer (Lord Jesus Christ, Son of God, have mercy on me, a sinner).

Once we enter the path of union, we will increase the rate at which we reap the fruits of our actions. The bulk of our karmic debt needs to be "burned off" before we can attain the goal. We have already discussed how actions performed with the idea that the fruits of the action are for the Lord rather than for personal gain burn off reactive momenta without creating new ones. Selfless actions can take various forms. Such service can be physical service, such as helping to build a shelter, or attending the sick; or economic service, such as relief work, feeding the poor, or helping the needy find a job. A higher form of service is intellectual service, such as teaching skills, general knowledge, morality, and spiritual philosophy. The highest form of service is spiritual service—performing spiritual practices. Such practices help us attain happiness, wisdom, empathy, and eventual unity. Naturally, such persons become a model for others to follow, leading others to the path of unity. Unlike physical forms of service, the effects of intellectual and spiritual service can be permanent in nature. All forms of true service weaken the grip of ego and bring one closer to the goal of unity. Hence service advances one's personal growth and has the secondary benefit of helping the persons served. In addition, by serving others out of the goodness of our hearts, without the need for even recognition or thanks, we will set an example that others will want to emulate, and they will be drawn to the spiritual path.

Yogic practices

As mentioned before yoga means "union." In the West, yoga is often confused with the postures or asanas of hatha yoga. However, before this third leg in Patanjali's eight-limbed path of yoga come the moral guidelines (*yamas* and *niyamas*) that are the moral underpinnings of this system of unification.[1] Yogic practices are designed to increase one's knowledge of self and thereby one's knowledge of God. A side effect is the possible development of psychic powers. There is the possibility that students of yoga may develop such powers and fall under their spell, using them for expanding their ego and controlling others. For this reason, yogic masters in the past would not take a student unless they felt the aspirant strictly followed moral principles. This is probably not as much of an issue as it once was because today most people coming to the spiritual path are doing so for attaining enlightenment and not to acquire psychic powers.

The yoga asanas are useful for calming, stretching, and vitalizing the body and glands. Their greatest value is in helping to attain a healthy body and calm mind so that one can better perform meditation. Besides asanas, hatha yoga prescribes a lacto-vegetarian diet free of tobacco, alcohol, and drugs. Such a diet is beneficial for the mind and body.

Yoga prescribes breath control or *pranayama* for purifying the psychic body. The psychic body is believed to contain psychic nerve channels (*nadiis*) and plexuses (chakras). It is believed that *pranayama* helps purify these psychic structures, allowing the latent spiritual energy (kundalini) that is aroused by spiritual practices to pass through the higher chakras and eventually reach the seat of consciousness at the top of the head (termed the "thousand-petaled lotus"). All of these earlier steps on the yogic path are designed to prepare the body and mind for contemplation of God—what may be termed meditation.

Meditation

Of all the various spiritual practices, meditation is probably the most powerful technique for achieving union. The goal of meditation is to achieve cosmic ideation. Many meditation practices have evolved over the ages for developing an awakened state of consciousness. Meditation practices are designed to draw one nearer to God by focusing the mind on God instead of the outside world. To become fully absorbed in God, the mind must become one-pointed or concentrated. However, concentrating the mind is not easy—it tends to jump around in an uncontrollable manner. In fact, the great sage Ramakrishna once likened the mind to a mad monkey. Then he corrected himself and said, "no, more like a mad monkey bitten by a scorpion."

Meditation practices probably had their origin about fifteen thousand years ago in what is now southern India. The practices gradually became incorporated into the various mystical traditions of the East, including yoga, Tantra, Buddhism, Hinduism, Taoism, and Sufism. Spiritual practices were probably first introduced to the West as early as ancient Egypt, whose religion appears to have been influenced by Asian mysticism. However, it was not until the end of the nineteenth century when Swami Vivekananda visited the United States that meditation practices became better known in the US. He was followed by numerous other teachers of Eastern mysticism, especially during the 1960–70s when there was an explosion of interest in all things Eastern. This "new age" movement was partly fueled by the use of psychedelic drugs (e.g. LSD), which were reputed to throw open the "doors of perception" and for a short time produce mystical-like experiences.

Patanjali broke meditation down into three phases. The first (*pratyahara*) involves withdrawing the mind from the activities of the sense and motor organs, which allows it to focus internally (*dharana*), after which the mind can attempt to associate with Cosmic Consciousness (*dhyana*).

In the broadest sense, meditation is any practice in which the ego is set aside, causing one's sense of doership to disappear as the mind becomes detached from bodily sensations. In the process, the person becomes lost in what they are doing, whether it is playing music or becoming fully engrossed in trying to solve a problem. When practiced for the purpose of spiritual growth the only difference is that the object of concentration is Cosmic Consciousness. However, Cosmic Consciousness is purely subjective and for this reason is sometimes called the "Supreme Subjectivity." This creates a difficult problem for anyone trying to practice meditation for the purpose of self-realization. The mind has a natural tendency to want to focus on something objective, but if the goal of spiritual practice is to transcend or extinguish the mind, then it has to become absorbed in "Supreme Subjectivity." Because of this conundrum, spiritual teachers have developed different techniques that aid a spiritual aspirant in their quest for limitlessness and divine bliss.

By simply sitting comfortably and watching the breath, the mind is calmed. This technique is perhaps the simplest form of meditation. It is the basis for mindfulness meditation. In this practice, the focus of the mind shifts from attention at its surface, which is in a constant dance of change, to a deeper experience of silence and calmness. Mindfulness means being aware of the observer in us. To perform simple mindfulness meditation, one sits comfortably in a place free of distractions, closes one's eyes, and lets the mind focus on the breath. Trying to remain in the here and now, any thought, image, or sensation is allowed to freely come and go without trying to push it out of the mind or paying any attention to it. If distractions come, then the mind is brought back to its focus on the breath.

In a recent study, even this simple meditation practice was shown to cause significant changes in the brain. Anatomical magnetic resonance images (MRI) were obtained from sixteen healthy, meditation-naïve participants before and after they underwent an eight-week mindfulness meditation program. Changes in gray matter density were measured and compared with a control group of seventeen individuals. The study found that an average of twenty-seven minutes of daily practice of mindfulness meditation produced a significant boost in gray matter density, specifically in the hippocampus, the area of the brain in which self-awareness, compassion, and introspection are associated. Furthermore, this boost of gray matter in the hippocampus was directly correlated to a decreased gray matter density in the amygdala—an area of the brain known to be instrumental in regulating anxiety and stress responses. In contrast, the control group did not experience changes in either region of the brain, thus ruling out the possibility that the changes observed were due to the passage of time.[2]

Another effective method of meditation involves using a mantra. In the West, the word is often used to mean a frequently repeated word, phrase, or slogan—especially to advocate something. For example, "Going Green is the mantra of the environmental movement." However, the word *mantra* comes from Sanskrit and literally means, "the quality that liberates the mind." The best-known mantra is probably "Om," but a mantra designed for meditation must be endowed with strong spiritual vibrations and a deep spiritual meaning—often a name of God. The use of a mantra for meditation takes advantage of two important qualities of mind. First, that the mind requires an object—simply trying to empty the mind of any thought or impulse is very difficult. A mantra serves as the object of concentration, and since it has a subtle vibration and since the mind takes on the qualities of the object of its attention, the mind becomes subtle. Secondly, the mind can only think of one thing at a time—if it can focus on the mantra then other thoughts and sensations cannot enter. Normally people experience a lot of activity in their mind. Such mental activity is actually unilateral (one thought at a time), much like the fast-moving frames of a motion picture. It just seems like the mind is thinking and experiencing multiple things simultaneously, but similar to a motion picture, what appears continuous is in fact made up of individual snapshots with little gaps in between.

A mantra is normally repeated silently although it may be sung aloud. The latter practice is known as *kirtan*. The mind is naturally attracted to the sweet vibration of the mantra, and ideally, it is drawn away from the chaos of the external world into a deep state of inner concentration and peace. A mantra may be considered a spiritual tool that is able to bridge the void between the objective level of existence (mind) and subjective reality—Cosmic Consciousness. It is best to learn meditation from a trained teacher. No single mantra is suited for everyone.

Besides meditation with a mantra, numerous other spiritual practices have been developed and used successfully by individuals for centuries.

Since every person is different, there can no more be a single method to attain unity with God then there can be a single diet that is good for everyone.

For example, Buddhists sometimes try to meditate on nothingness. In this practice, the meditator tries to empty their mind and enter the void or a state where the mind is extinguished (nirvana). This is a difficult meditation technique because the mind constantly tries to be engaged in some activity, be it thinking, feeling, or visualizing things. Thus it is usually done by more advanced practitioners, but it can be valuable in calming the mind and progressively clearing it of preconditioned beliefs and assumptions about life.

In Zen Buddhism (zazen), a student may be presented with an unsolvable, paradoxical statement or problem known as a koan. The koan is used as an aid to meditation and a tool for gaining spiritual awakening. An example is, "what is the sound of one hand clapping." The problem exists only in the mind. The minute the mind is suspended and sees through the illusion of an imaginary and separate self, the problem is solved. Separateness exists only in the mind—it is not real. The hand and the sound it makes are the same.

Visualization techniques are used in both Tantra yoga and Buddhist Tantra (*Vajrayana*). The form selected for ideation is normally that of the guru or a deity. Guru means "the one who dispels darkness." The guru has an important function in Tantra as not only the aspirant's guide on the spiritual path but also as the embodiment of Spirit. Of course, the Supreme Entity cannot be objectified, but for the aspirant the guru can represent the crystallized form of their higher self—a door to God. The spiritual aspirant (*sadhaka*) finds it difficult to visualize the subtle form, but in the process of trying to visualize it, the mind becomes subtler. This form of meditation (*dhyana*) helps the spiritual aspirant develop deep love for God.

Scientific studies have shown that meditation lowers blood pressure, lowers the level of blood lactate—reducing anxiety attacks, decreases tension-related pain and tension headaches, relieves ulcers, asthma, insomnia, muscle and joint problems, increases serotonin production, thus improving mood, and brings the brainwave pattern into an alpha state consistent with peacefulness, healing, and pain relief. Meditators report that they have increased energy, require less sleep, thus adding useful hours to the day, feel less stress, have improved metabolism, and can more easily lose weight. They feel more connected to other people and living things, and are simply happier than before they started meditating.

Other reported mental benefits of regular meditation include decreased anxiety, improved emotional stability, increased creativity, development of intuition, clarity and peace of mind, increased self-confidence, self-awareness, optimism, more harmonious relationships with friends, family and colleagues, sharper mind with improved focus, and improved emotional steadiness and harmony.

Looking at this list of physical and mental benefits of meditation might lead one to think that it is a panacea for all that ails humankind. This is possibly correct. However, the spiritual benefits of meditation are the most valuable and constitute the true purpose for the practice. These include a personal transformation, knowledge of who you truly are, gaining a deeper sense of purpose and a happier, more fulfilling life, attaining indescribable bliss, becoming free from the endless cycle of birth and death, and finally achieving self-realization or enlightenment (unity).

In the final analysis, all forms of spiritual meditation have a single purpose—to direct the mind away from the crude world of the sense and motor organs toward God. In the process, the unit "I" associated with ego becomes associated with the Cosmic "I"—an egoless state. The resulting mental expansion is accompanied by a new understanding of the meaning and purpose of existence and is accompanied by bliss along with greater love for God.

The final frontier—love of God

Intellectual knowledge is known as *jinana* in Sanskrit. One can have tremendous book knowledge about something but it is not the same as knowing or experiencing it. For example, a person may read a dozen books and study maps of Paris before embarking on a trip there, but would this measure up to their later experience of the city? Obviously not. There is no comparison between intellectual knowledge and experience, and this is especially true for spiritual experiences. Simply knowing in your mind that you are God does not create the feeling that you are God. Therefore, intellectual knowledge has limited benefit for the spiritual aspirant. Its greatest value lies in pointing the person toward the path and getting them started performing spiritual practices. At worst, it can create egoism because the person believes they are expert in something, which can be an obstacle for advancing on the path.

All the great sages have taught that God is best known through love. Spiritual practices are designed to unlock our unconditional love for God. Sanskrit has a specific word for devotion or love of God—bhakti; and the highest form of yoga is known as bhakti yoga.

In the first stage of bhakti, the spiritual aspirant feels that his or her Supreme Father is always with them. One feels God's existence everywhere, within and without—i.e. within the mind and in the physical world. One feels that whatever one does, whatever one is thinking, is being witnessed by God—nothing can be done secretly.

In the next stage of bhakti, the spiritual aspirant develops unconditional love for God and feels blissful. However, they love God not for their own bliss but because they want to give God bliss through their love and selfless service to all of God's created beings and objects.

The final stage of bhakti is called *kevala* bhakti. *Kevala* means "only," so in this type of bhakti there is only God—the spiritual aspirant and their Lord are one and the same. The aspirant enjoys the indescribable supreme bliss (ananda) from entering into the supreme ocean of Cosmic Consciousness. In this state, there is not even a touch of duality and the devotee forgets themselves and all their little predicaments—their unity with God is complete. This state could last for a moment or for the rest of one's life.

The path to unity is paved by spiritual practices, and the goal of unity is obtained when there is nothing in the mind of the devotee but their love for God. I hope it is clear to the reader that nothing in life can compare to being fully ensconced in the ecstasy of oneness with God.

Notes

Introduction

1.　Capitalization is used throughout this book for words referring to a singular, transcendental entity, quality, or thing. For example, Cosmic Mind, the Mind of God, or One Mind. Sanskrit terms are sometimes used parenthetically because they are more specific and offer a better description than the English equivalent.
2.　Ian Stevenson, *Children Who Remember Previous Lives: A Question of Reincarnation* (Jefferson, NC: McFarland, 2001) 30.

Chapter 1

1.　Anandamurti, *Namah Shivaya Shantaya* (Purulia, India: Ananda Marga Publications, 1982), 8.

Chapter 3

1.　Wisdom of Solomon 8:19-20.
2.　Wisdom of Solomon 8:19-20.
3.　Malachi 4:5-6.
4.　Elizabeth C. Prophet and Erin L. Prophet, *Reincarnation: The Missing Link in Christianity*, (Gardiner, MT: Summit Pub., 1997) 79.

Chapter 4

1.　Herbert B. Puryear, *Why Jesus Taught Reincarnation* (Scottsdale, AZ: New Paradigm Press, 1993) xii.
2.　See not only Puryear (above), but also: Elizabeth C. Prophet and Erin L. Prophet, *Reincarnation: The Missing Link in Christianity* (Gardiner, MT: Summit Pubs. 1997); Quincy Howe Jr., *Reincarna-*

tion for the Christian (Wheaton, IL: Theosophical Pub. 1974); and Geddes MacGregor, *Reincarnation in Christianity: A New Vision of the Role of Rebirth in Christian Thought* (Wheaton, IL: Theosophical Pub. 1978).
3. Matthew 16:14, Mark 6:15, Luke 9:8.
4. Malachi 4:5-6.
5. Matthew 11:13-14.
6. Mark 1:2-4, Luke 1:17; 7:27.
7. Matthew 17:10-13.
8. Luke 1:17.
9. 2 Kings 1:8, Matthew 3:4.
10. Matthew 3:1.
11. 1 Kings 18:17, Matthew 14:3.
12. 1 Kings 8:40 and 19:1.
13. 2 Kings 2:11.
14. John 10:34.
15. John 10:30.
16. John 3:5-6
17. John 9:1-3, 7.
18. Genesis 9:6.
19. Exodus 21:12, 23-25.
20. Galatians 6:7, Matthew 26:52.
21. Galatians 6:5, 7-8.
22. Elaine Pagels, *The Gnostic Gospels* (NY: Random House, 1979).
23. Malachi 1:3.
24. John 1: 1-3.
25. John 14, 15:5.
26. Elizabeth C. Prophet, *The Lost Years of Jesus: Documentary Evidence of Jesus' 17-year Journey to the East*, (Gardiner, MT: Summit Pubs. 1984).

Part II

1. Noel Langley, *Edgar Cayce on Reincarnation*, (NY: Warner Books, 1967) 11.

Chapter 6

1. Ian Stevenson, *Children Who Remember Previous Lives: A Question of Reincarnation* (Jefferson, NC: McFarland, 2001).
2. Ian Stevenson, *Reincarnation and Biology: A Contribution to the Etiology of Birthmarks and Birth Defects* (NY: Praeger, 1997).
3. Ian Stevenson, *Telepathic Impressions: A Review and Report of 35 New Cases* (Charlottesville, VA: University Press, 1970) 7.
4. Tom Shroder, *Old Souls: The Scientific Search for Proof of Past Lives* (NY: Simon & Shuster, 1999).
5. Jesse Bering, "Ian Stevenson's Case for the Afterlife: Are We 'Skeptics' Really Just Cynics?" *Scientific American Blog Network*, (Nov. 2, 2013).

Chapter 7

1. Jim. B. Tucker, *Life Before Life: Children's Memories of Previous Lives* (NY: St. Martin's Griffin, 2005) 1-3.
2. https://www.youtube.com/watch?v=0ppZlC2jwag.
3. NBC Nightly News with Lester Holt on March 20, 2015; https://www.nbcnews.com/nightly-news/video/boy-remembers-amazing-details-of-past-life-as-hollywood-actor-416079939861.
4. Jim B. Tucker, *Return to Life: Extraordinary Cases of Children who Remember Past Lives* (NY: St. Martin's, 2013) 88-119.

Chapter 8

1. Raymond A. Moody, Jr. and Paul Perry, *Coming Back: a Psychiatrist Explores Past-life Journeys* (NY: Bantam, 1991).
2. Brian L. Weiss, *Many Lives, Many Masters: The True Story of a Prominent Psychiatrist, His Young Patient, and the Past-Life Therapy That Changed Both Their* Lives (NY: Simon & Shuster, 1988).
3. Carol Bowman, *Children's Past Lives: How Past Life Memories Affect Your Child* (NY: Bantam, 1997).
4. http://www.reincarnationforum.com.
5. Peter Fenwick and Elizabeth Fenwick, *Past Lives: An Investigation into Reincarnation Memories* (NY: Berkley, 1999).
6. Shirley MacLaine, *Out on a Limb* (NY: Bantam, 1983).

7. Peter Fenwick and Elizabeth Fenwick, *Past Lives: An Investigation into Reincarnation Memories*, 82-91.

Chapter 9

1. Norman C. McClelland, *Encyclopedia of Reincarnation and Karma* (Jefferson, NC: McFarland, 2010) 58.

Chapter 10

1. Michael J. Behe, *Darwin's Black Box: The Biochemical Challenge to Evolution* (NY: Free Press, 1996).
2. Thomas Nagel, *Mind and Cosmos: Why the Materialist Neo-Darwinian Conception of Nature is Almost Certainly False* (Oxford: Oxford University Press, 2012).
3. Then intelligent life should be very common in the universe, right? Probably not. The universe contains countless stars, many of which have planetary systems, but only a few of these would have a solid crust and be neither too warm nor too cold for liquid water. There are probably only a small percentage of those planets where life developed and had favorable conditions long enough for the evolution of highly intelligent living beings. This, along with the vast distance between stars within a galaxy, makes it unlikely that we will ever make physical contact with intelligent beings from another world.
4. Rupert Sheldrake, *A New Science of Life: The Hypothesis of Morphic Resonance* (Los Angeles: J. P. Tarcher, 1981).

Chapter 11

1. Max Planck quoted in *Accent Magazine*, "The Hidden World of Mind," Oct. 1972, New Delhi, India.
2. Henry Margenau, *The Miracle of Existence* (Woodbridge, CT: Ox Bow, 1984) 96.
3. Edward F. Kelly and Emily W. Kelly, *Irreducible Mind: Toward a Psychology for the 21st Century* (Lanham, MD: Rowman & Littlefield, 2007).
4. Ibid. 45.
5. Von der Malsburg, "Binding in Models of Perception and Brain Function," *Current Opinion in Neurobiology*, **5**, (1995): 520-

526.
6.   Richard Gerrig and Phillip Zimbardo, *Psychology and Life*, 20th Ed. (Essex England: Pearson Education, 2014).
7.   Rick Hanson and Richard Mendius, *Buddha's Brain: The Practical Neuroscience of Happiness, Love and Wisdom* (Oakland, CA, New Harbinger, 2009) 7
8.   J. R. Searle, "Minds, Brains, and Programs," *Behavioral and Brain Sciences*, **3**, (1980): 417-24.

Chapter 12

1.   Von der Malsburg, "Binding in Models of Perception and Brain Function," *Current Opinion in Neurobiology*, **5**, (1995): 520-526.
2.   R.L. Moody, "Bodily Changes during Abreaction," *Lancet*, 1, (1948): 964.
3.   Edward F. Kelly and Emily W. Kelly, *Irreducible Mind: Toward a Psychology for the 21st Century*, 129.
4.   Ibid. 216.
5.   L.K. Kothari, Arun Bordia, and O. P. Gupta, "The Yogic Claim of Voluntary Control over the Heart Beat: an Unusual Demonstration." *J. American Heart Assoc.*, **86**, no. 2, (1973): 284.
6.   Ian Stevenson, *Telepathic Impressions: A Review and Report of 35 New Cases* (Charlottesville, VA: University Press, 1970).
7.   Edward F. Kelly and Emily W. Kelly, *Irreducible Mind: Toward a Psychology for the 21st Century*, 172-3.
8.   Larry Dossey, M.D. *Healing Words: The Power of Prayer and the Practice of Medicine* (NY: HarperCollins, 1993).

Chapter 13

1.   Robert Ullman and Judyth Reichenberg-Ullman, *Mystics, Masters, Saints, and Sages: Stories of Enlightenment* (Berkeley, CA: Conari Press, 2001) 37.
2.   Gopi Krishna, *Living with Kundalini* (Boston: Shambhala, 1993).

Chapter 14

1. H. Hart, "ESP Projection: Spontaneous Cases and the Experimental Method," *Journal of the American Society for Psychical Research*, **48**, (1954): 121-46.
2. Sam Parnia, *Erasing Death: The Science that is Rewriting the Boundaries between Life and Death* (NY: HarperCollins, 2014).
3. Robert Bruce and Brian Mercer, *Mastering Astral Projection: 90-day Guide to Mastering Out-of-body Experience* (St. Paul, MN: Llewellyn, 2004).
4. These include the Monroe Institute's Nancy Penn Center in Virginia, the Center for Higher Studies of Consciousness in Brazil, the International Academy of Consciousness in southern Portugal, which features the Projectarium, a spherical structure used exclusively for practice and research on out-of-body experience, and Olaf Blanke's Laboratory of Cognitive Neuroscience in Switzerland.
5. Raymond A. Moody, Jr., *Life after Life: The Investigation of a Phenomenon— Survival of Bodily Death* (NY: HarperCollins, 1975).
6. Jeffrey Long and Paul Perry, *Evidence of the Afterlife: The Science of Near-Death Experiences* (NY: HarperCollins, 2010).
7. Edward F. Kelly and Emily W. Kelly, *Irreducible Mind: Toward a Psychology for the 21st Century*, 419.
8. Raymond A. Moody, *Intelligence Squared* debate entitled "Death is not Final;" http://intelligencesquaredus.org/images/debates/past/transcripts/050714%20Death%20Not%20Final.pdf.

Chapter 15

1. David Bohm, "A Suggested Interpretation of the Quantum Theory in Terms of Hidden Variables". *Physical Reviews* **35** (2): (January 15, 1952): 188.
2. Literally an analysis of analyses. The aim is to derive a statistical analysis from a pool of similar studies. This allows for combination of information leading to higher statistical certainty than is possible from any individual study.
3. Dean Radin, *The Conscious Universe: The Scientific Truth of Psychic Phenomena* (NY: HarperCollins, 1997) 79-80.
4. Ibid. 87-89.
5. Guy L. Playfair, *Twin Telepathy* (Guildford, UK: White Crow, 2011).
6. Dean Radin, *The Conscious Universe: The Scientific Truth of*

Psychic Phenomena, 99-100.
7.   J. G. Pratt, et al, *Extrasensory Perception after Sixty Years* (Boston: Bruce Humphries, 1966) 42.
8.   Dean Radin, *The Conscious Universe: The Scientific Truth of Psychic Phenomena*, 120.
9.   Ibid. 126-130.
10.   Ibid. 143.
11.   Hardware RNGs are preferred and may employ radioactive decay or electronic noise. Noise spikes or decay signals are measured against an electronic oscillator such as a quartz crystal to produce thousands of spikes per second, which can be translated into a truly random sequence of zeros and ones that are accurately recorded by computer.
12.   Dean Radin, *The Conscious Universe: The Scientific Truth of Psychic Phenomena*, 151.
13.   Sean Carrol, *Discover Magazine Blogs*, http://blogs.discovermagazine.com/cosmicvariance/2008/02/14/american-association-for-the-advancement-of-pseudoscience/#.XInSEqBKhPY.
14.   Dean Radin, *The Conscious Universe: The Scientific Truth of Psychic Phenomena*, 81-2.
15.   Ibid. 80-1.
16.   Ibid. 100.
17.   A gas discharge tube is a bulb or tube (usually glass) with two or more electrodes that has been evacuated and filled with a gas or gas mixture, usually at somewhat less than atmospheric pressure. As the voltage applied across the electrodes is increased, there comes a point called the breakdown voltage at which ionization of the gas will initiate an avalanche process that spreads through the tube. The voltage at which breakdown occurs depends on various factors, such as gas composition, pressure, spacing of the electrodes, etc.
18.   William A. Tiller., Walter Dibble, and Michael Kohane, *Conscious Acts of Creation: The Emergence of a New Physics* (Walnut Creek, CA: Pavior Publishing, 2001) 1-13.
19.   B. Dunne and R. Jahn, "Experiments in Remote Human/Machine Interaction, *Journal of Statistics Edu.*, **6** (1992): 311-32.
20.   http://pearlab.icrl.org/experiments.html.
21.   H. E. Puthoff and R. Targ, "A Perceptual Channel for Information Transfer over Kilometer Distances: Historical Perspective and Recent Research," *Proceeding of the Institute of Electrical and Elec-*

tronic Engineers, **64** (1976).
22.   Dean Radin, *The Conscious Universe: The Scientific Truth of Psychic Phenomena*, 105.
23.   Dean Radin, Entangled Minds: *Extrasensory Experiences in a Quantum Reality* (NY: Paraview, 2006) 195-202.
24.   Dean Radin, *Entangled Minds: Extrasensory Experiences in a Quantum Reality*, 136-141.
25.   Gary E. Schwartz, *The Sacred Promise: How Science is Discovering Spirit's Collaboration with us in our Daily Lives* (NY: Atria Books, 2011)
26.   Dean Radin, *Supernormal: Science, Yoga, and the Evidence for Extraordinary Psychic Abilities* (NY: Chopra Books, 2013).
27.   Maryann Mott, *National Geographic News*. January 4, 2005.
28.   Don Oldenburg, *Washington Post*. January 8, 2005.
29.   Maryann Mott, *National Geographic News*. November 11, 2003.
30.   Larry Dossey, *Recovering the Soul: a Scientific and Spiritual Search* (NY: Bantam, 1989), 113.
31.   Dean Radin, *The Conscious Universe: The Scientific Truth of Psychic Phenomena*, 275.

Chapter 16

1.   Neil Savage, "Seeking Materials to Send Unbreakable Codes," *Chemical & Engineering News*. September 11, 2017.
2.   Avery Thompson, "How Quantum Teleportation Actually Works," *Popular Mechanics*. March 16, 2017.
3.   David Mermin, "Extreme Quantum Entanglement in a Superposition of Macroscopically Distinct States." *Physical Review Letters*, **65** (1990).

Chapter 17

1.   Elizabeth C. Prophet and Erin L. Prophet, *Reincarnation: The Missing Link in Christianity*, (Gardiner, MT: Summit Pubs. 1997) 278.
2.   Ibid. 278-9.
3.   Deepak Chopra, *The Third Jesus* (NY: Harmony Books, 2008) 161.
4.   Herbert B. Puryear, *Why Jesus Taught Reincarnation* (Scotts-

dale, AZ: New Paradigm Press, 1993) 214-17.
5.   Luke 17:21.

Chapter 18

1.   Carl G. Jung, *Modern Man in Search of a Soul* (NY: Harcourt Brace Jovanovich, 1933).
2.   Deepak Chopra, *The Way of the Wizard* (NY: Harmony Books, 1995) 161.
3.   Luke 18:17
4.   Mark 10:24, Matthew 5:44.
5.   Matthew 5:39.

Chapter 19

1.   Yamas (observances):
Ahimsa: Non-violence, non-harming (do no intentional harm in thought, word, or actions).
Satya: Truthfulness or honesty (use of speech in the spirit of welfare).
Asteya: Non-stealing.
Brahmacharya: Walking in awareness of the highest reality, remembering all beings and things are the Divine.
Aparigraha: Non-possessiveness, the non-hoarding of wealth that is superfluous to our needs.
        Niyamas (self-controls):
Shaocha: Purity of mind and body, cleanliness. This is both internal and external. How best to treat our bodies and our energies and how we treat our environment.
Santosha: Contentment, mental ease. Be content with what you attain. Accept where you are. The quality of contentment leads to inner peace.
Tapah: Disclipline, to undergo hardship when acting in the service of others without expecting anything in return.
Svadhyaya: A clear understanding of a spiritual subject. Also the study of one's self; careful self-observation; turning inward.
Ishvara Pranidhana: To make the Supreme Consciousness the goal of our life. Surrender to God.
2.   Britta K. Hölzel, James Carmody, Mark Vangel, Christina Congleton, Sita M. Yerramsetti, Tim Gard, and Sara W. Lazara, "Mindful-

ness Practice Leads to Increases in Regional Brain Gray Matter Density," *Psychiatry Research: Neuroimaging*, **191**, no. 1 (2011): 36–43.

Index

A

acharyas 158
acupuncture 129
Adam 19, 31, 44
Adams, Robert 106
Advaita Vedanta 9
agama 8
Alex (the African grey parrot) 83
Ambrose 19
Ammonius Saccas 27
amnesia 119
amygdala 174
ananda 11, 107, 177
Anandamurti 158, 159, 160, 178
anatomical convergence 94
Andromeda Galaxy 110
animal instincts 4, 80, 90
anomalies 54, 76
Aquinas, St. Thomas 32
arrow of time 148
asanas 156, 171
astral projection 116
astrology 129
Atlantic salmon 81
atman 1, 9, 12, 38, 155
autism 72
avadhutas 159
AWARE studies 116
Ayurveda 8

B

ba 14
Babylon 157
bacterial flagellum 79
Baha'u'llah 105
Behe, Michael 85, 181
Bernstein, Morey 51
Besant, Annie 36
Bhagavad Gita 7, 156
bhakti 177
bhakti yoga 177
Big Bang 39, 144, 149
biological reincarnation 84
birth defects 3, 50, 54, 55, 58, 59
birthmarks 3, 50, 54, 55
bits 128, 134, 142
black hole 109, 111
Blake, William 3, 52
block time 112
bodiless mind 41, 42, 45
Bohm, David 183
Bohr, Neils 150
bowerbird 81
Bowman, Carol 66, 67, 180
Brahma 9, 155
Buddha 8, 10, 11, 12, 44, 105, 135, 156, 182
Buddhism 7, 9, 10, 39, 156, 172, 175
Buddhist Tantra 175

C

Cambrian Period 80
Cardiac, Jean-Louis 73
Carrol, Sean 128, 129, 184

Catholic 26, 137
Cayce, Edgar 49, 50, 179
chakras 171
child prodigies 71, 73
chiropractic 129
Chopra, Deepak 158, 164, 185, 186
Christ 26, 28, 34, 101, 170
Christenson, Patrick 54
clairvoyance 116, 123, 125, 131
Clement of Alexandria 19
collapse of the wave function 93, 147, 148
collective unconscious 132, 165
Cosmic Consciousness 1, 9, 11, 37, 38, 39, 45, 84, 105, 113, 150, 156, 159, 165, 172, 173, 174, 177
Cosmic Entity 8
Cosmic Mind 38, 39, 40, 41, 46, 77, 82, 86, 93, 132, 149, 159, 165, 178
Council of Constantinople 33
cryptomnesia 65, 68
cyanobacteria 77
cyclotron 109

D

damnation 28, 31, 43, 44
Darwin, Charles 78, 80, 85, 181
Dawkins, Richard 79
Dead Sea Scrolls 18, 26
déjà vu 65
Devi, Shanti 62, 63, 64
dharana 156, 172
dharma 9, 159
dhyana 156, 172, 175
diffraction pattern 146
Division of Perceptual Studies 52, 58, 61
DNA 76, 79, 80, 81, 86, 132
dogmas 167
dogs 53, 85, 136
Dossey, Larry 103, 182, 185
dreams 42, 60, 114, 121
dualism 9
dual-slit experiment 144, 145

E

Earth 77, 78, 108, 110, 111, 112, 137, 138
Eastern mysticism 13, 14, 172
Eastern religions 1, 2, 4, 7, 9, 32
East Hebrides 59
Eddington, Arthur Stanley 109
Egypt 19, 172
Eight-Fold Path 11
Einstein, Albert 75, 108, 141, 150
EKG 102
electromagnetic energy 39
Elijah 19, 22, 23, 24
Emerson, Ralph Waldo 3, 62
endorphins 120
End Time 2
engrams 96
enkephalins 120
enlightenment 11, 105, 153, 165, 166, 168, 171, 176
entanglement 141, 142, 143, 144
epiphenomenalism 122
Er 16
espionage 134
Essenes 18, 22
eternal now 108
Eve 19, 31, 44
evolution of species 4, 76, 80
ex nihilo 31, 32

F

false pregnancy 101
Fenwick, Peter and Elizabeth 69, 180, 181
four-dimensional sight 112
Four Noble Truths 11
Franklin, Benjamin 3, 71
Franklin, Sir John 68
fundamental factors 39

G

Galileo 75, 137
ganzfeld telepathy experiments 130
Gehrig, Lou 74
general theory of relativity 108, 109
Genesis 25, 179
genius 3, 28, 50, 71, 75
gilgul 20, 157
Global Consciousness Project 134
gnosis 26, 27, 157
Gnostic Christians 2, 26, 27, 157
Gnostic Gospels 22, 26, 27, 179
Gnosticism 27, 157
Golden Rule 25, 36
Gorilla Sign Language 83
Gospel of Thomas 155
GPS system 110
gravitational waves 110
gravity 109, 111, 150
guru v, 105, 168, 175

H

Haicheng, China 136
hallucination 115
Hammons, Ryan 60
Hanumant 55
Hasidic Judaism 20
hatha yoga 171
Haupt, Christian 74
Hebrew Bible 17
Heisenberg, Werner 140
hell 16, 43, 44
hemophiliacs 79
higher self 175
Himis manuscript 33
Hinduism 7, 9, 10, 14, 39, 105, 156, 172
hippocampus 173
homeopathy 129
homing ability 137
homunculus 92, 97
Huxley, Aldous 155
hyperlexia 72
hyperthymesia 96
hypnosis 42, 49, 50, 51, 65, 66, 68, 69, 95, 100, 101, 105
hypnotic regression 3, 42, 65, 67

I

ideology 35, 37, 38, 39, 41, 43, 45, 80, 82
immortality 15, 22, 25, 31, 35, 164
immune system 79, 80, 100
intentionality 92
interference pattern 145
irreducibly complex 79
Islam 2, 4

J

Jacob and Esau 28
Jainism 7
Jeans, James 114

INDEX

Jeremiah 18, 22
Jesus Prayer 170
Jews 17, 21, 22, 26
jinana 176
John the Baptist 22, 23
Jones, Bobby 61
Josephus 17, 18
Judaism 2, 4, 17, 19, 20, 24, 105, 157
Jung, Carl 3, 162, 163, 186

K

ka 13, 14
karmic burden 44, 45, 160, 165
Kelly, Edward F. 93, 181, 182, 183
ketamine 121
Kingdom of God 24, 158, 165
kirtan 174
koan 175
Krishna 7, 106, 107, 119, 156, 182
Kuhlmann-Wilsdorf, Doris 56
kundalini 106, 107, 171

L

lacto-vegetarian diet 171
Lao Tzu 10, 105, 156
law of conservation of energy 93
law of karma 1, 25, 36, 37, 41, 43, 44, 45, 50
Leininger, James 64
Lewis-Clack, Rex 72
liberation 8, 9, 12, 105, 153, 169
LIGO observatories 110
limitlessness 9, 159, 162, 165, 173
Logos 19

Long, Jeffrey 118, 183
LSD 115, 121, 172
Luke 22, 26, 27, 179, 186

M

Macaulay, Cameron 59
MacLaine, Shirley 3, 69, 180
madhuvidya 166
Mahabharata 156
Malachi 19, 22, 23, 178, 179
mantra 174
Margenau, Henry 93, 181
Mark 3, 22, 26, 27, 179, 186
Mars 141
Martyn, Marty 60, 61
material Darwinism 78, 80, 85
materialism 113, 138, 139, 167
materialist ideology 37, 39
materialist worldview 77, 89, 149
Matthew 22, 26, 27, 179, 186
maya 9, 156
McConnell, John 59
meditation 8, 11, 42, 43, 68, 100, 104, 106, 116, 135, 156, 158, 159, 160, 164, 171, 172, 173, 174, 175, 176
mediums 135
Mental Internet 132
Messiah 19, 22
meta-analysis 127, 128, 130, 131
metaphysics 138
metempsychosis 14, 15
meteorites 138
Michelson, Albert 108
midlife crisis 163
mind-body unity 100
mindfulness meditation 173

mind-stuff 38, 39, 40, 82
Mishnah 17
Mohammed 105
moksha 9, 105
monarch butterflies 81
monistic philosophy 9
Moody, Raymond A. 64, 66, 99, 118, 121, 180, 182, 183
Moon 150
morality 11, 170
Morley, Edward 108
Moses 19, 105
Mozart 75
mudras 8
mukti 105
multicellular organisms 78, 82
Murphy, Bridey 51
Myers, F.W.H 124
mystical experience 90, 104, 105, 106, 108, 113

N

nadiis 171
Nagel, Thomas 80, 181
Nag Hammadi 26
natural selection 37, 80
Nelson, Roger 3, 134
Neoplatonists 2, 28
neuron 97
neuroscience 91, 122
new age 172
New Testament 19, 22, 23, 24, 25
Newton 36, 75
Newton, Isaac 36, 75
Nicaean Council 27
Nicaean Creed 29, 30
Nicodemus 24
nigama 8

nirvana 105, 153, 175
nirvikalpa samadhi 102
niyama 156
nocebo 100
noncerebral memory 97
Notovich, Nicolas 33

O

occult powers 8
Old Testament 18, 19, 22, 28, 44, 157
One Mind 93, 132, 178
ontology 37, 39, 89, 139
Origen 2, 19, 27, 28, 29, 31, 33
original sin 31, 44
Orpheus 14
Orphism 14, 15

P

panentheism 32, 33
Pang, Tristan 75
pantheism 32
Paramatman 1, 9, 14, 38
paramecium 82
paranormal phenomena 56, 129
parapsychology 123, 128
Parnia, Sam 116, 119, 183
Parvati 8
past-life regression 51, 65, 66, 67
Patanjali 156, 171, 172
path of unity 167, 170
Peek, Kim 72
perennial philosophy 155
Pharisees 18, 22
Philo 19
phobias 42, 65, 66
photon 110, 143, 145, 146
photosynthesis 77, 82

physicalism 138
placebo 89, 99, 100, 101
Planck, Max 92, 93, 181
planets 85, 106, 138, 181
Plato 3, 14, 15, 16, 19, 31, 113, 118
Platonism 16, 27, 30
polarization 144
Pope John Paul II 134
Prabhat Samgiita 159
prana 13
pranayama 156, 171
pratisaincara 159
Pratt, J.G. 125, 184
pratyahara 156, 172
prayer 103, 104, 169, 170
precognition 123, 126, 131
preexistence of the soul 17, 22, 27, 28, 30, 33
presentiment 126, 130
preya 9
Princess Diana 3, 134
Principe Island 109
probability 56, 85, 93, 134, 136, 144, 146, 147, 148
Procopius 33
Prophet, Elizabeth Clare 33, 34, 178, 179, 185
proteins 80, 81, 82
pseudoscience 128, 129, 138, 184
pseudo-skepticism 129
psi 123, 130, 131, 132, 133, 134, 139
psychic body 116, 171
psychic phenomena 90, 138, 139
psychic powers 135, 171
psychokinesis 123, 127, 131

psychoneuroimmunology 100
purgatory 18
Puryear, Herbert 178, 186
Pythagoras 3, 14, 15, 18, 19

Q

quanta 144, 146, 150
quantum communication 143
quantum computers 142
quantum mechanics 92, 141, 143, 144, 146, 148, 149, 150
quantum physics 142, 144
quantum teleportation 143
quantum wave function 147, 148

R

Rabbi Moses de Leon 19
Rabbi Simeon bar Yochai 19
Rabbi Yonassan Gershom 69
Radin, Dean 128, 135, 183, 184, 185
Ramakrishna 104, 172
Ram Dass 168
reactive momenta 40, 41, 43, 45, 160, 161, 163, 166, 167, 170
reciprocal apparitions 116
reductionism 138, 167
Reincarnation Forum 67
relativity theory 112, 149
religion 7, 9, 10, 29, 30, 46, 105, 118, 137, 167, 172
remote viewing 116, 123, 125, 134
resurrection 24, 26, 27, 31, 157, 158
RNA 80
robotics 142

S

Sadducees 17
sadhaka 175
sadhana 8, 160
Sagan, Carl 58
sage 8, 9, 156, 172
saincara 159
samadhi 11, 102, 105, 153, 156, 165
Sanskrit 1, 9, 40, 45, 156, 166, 174, 176, 177, 178
Sarkar, P.R. 158
Satan 44
satori 105
Satyamurti 102
savant 72, 73
Schrödinger, Erwin 147
Schrödinger's Cat 147
Schrödinger wave equation 147
Schwartz, Gary 135, 185
scientism 57
seeker 164, 165
seer 165, 168
selfless service 43, 46, 164, 166, 169, 177
self-realization 35, 38, 43, 105, 153, 159, 160, 163, 164, 173, 176
Shankara 9
Sheldrake, Rupert 76, 85, 181
Shiva 8, 105, 156
shreya 9
Shroder, Tom 56, 180
shunya 12, 156
Shunyavada 12
siddhis 116, 135
Sikhism 7
sin 31, 34, 43, 44
skin-writing 102
Socrates 3, 13, 14, 15
space-time 39, 108, 109, 110, 111, 112, 123, 149, 150
spatial nonlocality 144, 146, 148
Spirit 24, 25, 26, 31, 162, 163, 164, 166, 169, 175, 185
spiritual ideology 37, 38, 39, 41, 43, 80
spirituality 8, 39, 159
spiritual philosophy 37, 50, 170
spiritual union 1, 25, 27, 157, 158, 159
spiritual worldview 37, 43, 45, 89, 149
Sri Lanka 52, 53, 136
stars 106, 109, 138, 181
Star Trek 143
Stevenson, Ian 50, 51, 52, 53, 54, 55, 56, 57, 58, 61, 62, 69, 103, 178, 180, 182
St. Francis of Assisi 101
stigmata 90, 101
subconscious 105, 120
Sufism 39, 105, 172
sun 7, 16, 38, 77, 166
supercomputers 142
supernatural 131
superposition 93, 142, 147, 148
superpowers 135
Supreme Subjectivity 173
surrender 156, 161, 163, 169
Swann, Ingo 134

T

Talmud 17
Tammet, Daniel 73, 74
Tantra 8, 39, 159, 172, 175
Taoism 7, 10, 39, 172

Tao Te Ching 10
Taylor, Gus 59
telepathy 123, 124, 125, 130, 131
temporal nonlocality 144, 145, 146
Ten Commandments 43
Theodora 33
Tighe, Virginia 51
Tiller, William A. 133, 184
Tlingit Indians 2, 52
Tree of Knowledge 44
Truzzi, Marcello 129
tsunami 136
Tucker, Jim 51, 58, 59, 60, 61, 62, 180
turiya 107

U

unhappiness 41, 43, 163, 166
unit consciousness 1, 12, 38, 156
unit mind 1, 3, 40, 43, 45, 82, 84, 85, 86, 89
unity of self-awareness 94, 95
University of Virginia 51, 52, 58, 61
Upanishads 9, 155

V

Vatican 34
Vedanta 7, 9, 39
Vivekananda 172

W

wave function 93, 144, 147, 148, 149, 150
Weiss, Brian 66, 180

Wiltshire, Stephen 72
Wisdom of Solomon 18, 178

X

xenoglossy 69

Y

Y2K 134
Yala National Park 136
yama 156
yang 10
yin 10
yoga 39, 116, 156, 158, 159, 171, 172, 175, 177
Yoga Sutras 156

Z

Zen 175
Zohar 19, 20

www.ingramcontent.com/pod-product-compliance
Lightning Source LLC
Chambersburg PA
CBHW021102080526
44587CB00010B/340